Optimization:
A Simplified Approach

WILLIAM CONLEY
University of Wisconsin—Green Bay

Optimization: A Simplified Approach

new york / princeton

Dedicated to the
memory of my
grandfather
James Joyce

Typesetting by Backes Graphics

Printed in the United States
1 2 3 4 5 6 7 8 9 10

Library of Congress Cataloging in Publication Data

Conley, William, 1948-
Optimization: a simplified approach.

Bibliography: p.
Includes index.
1. Mathematical optimization–Data processing. 2. Basic (Computer program language) I. Title.
QA402.5.C643 519 81-8513
ISBN 0-89433-121-3 AACR2

Contents

Introduction

The idea of optimization is an attractive one. Generally, it means finding the best course of action, the best way of doing something. In this book we look at optimization from a mathematical point of view.

We will look at profit equations, cost equations, chemical yield equations, inventory models, and such problems as product mix, demand analysis, package design, transportation, medical research, blending, fixed cost, discounts for buying in quantity, financial planning, physics, engineering, and cargo loading. In each case, we describe a problem as a function to be maximized or minimized, subject to certain conditions or constraints. Then we solve the problem (find the optimum) by writing a program in the computer language, BASIC. In the course of the presentation, BASIC will be discussed and its use explained.

Most books on optimization rely heavily on calculus and linear programming, two techniques that were the way to go in the precomputer age and early in the computer age. These approaches are still recommended when appropriate. For nonlinear optimization problems of 1 to 100 variables, however, one should write a computer program to optimize the function. We shall do this. Because the technique presented in this text always works on these nonlinear problems, it frees the engineer, scientist, businessperson, or teacher to create and solve accurate nonlinear models instead of oversimplified linear models.

Taking this approach, we can not only get the right answer to our question, but for the first time we will be able to get the right answer to the *right question*.

Everything in the text is explained carefully and repeatedly. Depending on their background, readers can set their own pace.

Also note that the statistical distributions of integer programming problems (the graphs) that appear in this book are part of the statistical justification of Monte Carlo, focus search, exhaustive search, and multi-stage Monte Carlo integer–programming approaches to solving optimization problems. Chapter 10 is devoted to the statistical justification and demonstration of the validity of our approach. It can be read at any time.

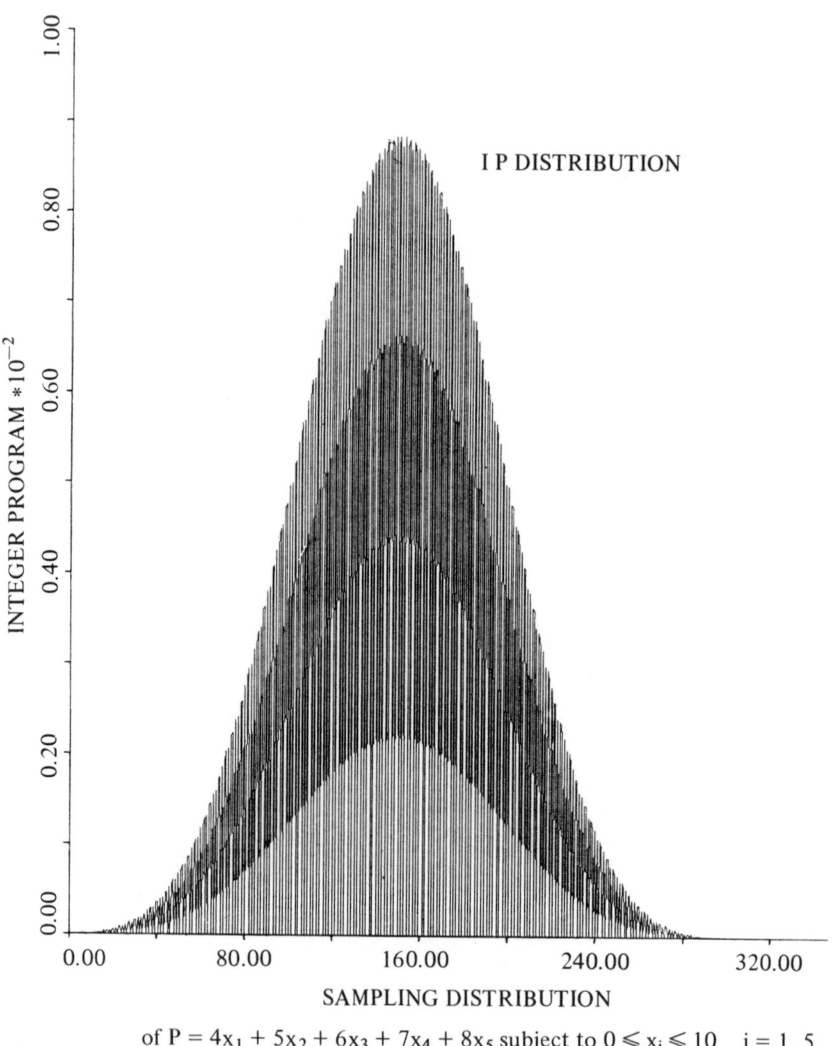

FIGURE 1.1 is a graph of all the feasible solutions of $P = 4x_1 + 5x_2 + 6x_3 + 7x_4 + 8x_5$ subject to $0 \leqslant x_i \leqslant 10, i = 1,5$. The maximum answer is at the extreme right (when $x_1 = x_2 = x_3 = x_4 = x_5 = 10$). The minimum answer is at the exteme left (when all x_i's $= 0$ and $p = 0$). The most frequently occurring answer is in the middle of the distribution (highest peak).

The graph is a representation of the pattern of occurrence of all $11^5 = 161{,}051$ answers. (We are allowing only whole number answers 0,0,0,0,0 through 10,10,10, 10,10.)

CHAPTER 1

BASIC and Optimization

Humans can do many things that computers cannot do. And humans can do almost everything that computers can do. So why have computers? There are many answers to this question, but the major reason to use computers is to harness and exploit their truly spectacular speed. In this work we are referring specifically to their computation speed.

Every problem in this book (including the one hundred exercises) can be worked out with pencil and paper. However, it will take a person (on the average) a few years per problem. But it will take the computer only several seconds or a few minutes.

With this general goal in mind, let us now look at the computer language BASIC and work on some examples.

BASIC consists of a series of statements numbered in an ascending order and entered into the computer either with a computer terminal (like a teletype typewriter) or on punched computer cards or some other means. (The programs in this book were typed in and run on a computer terminal.) The computer runs or executes the programs sequentially from top to bottom, unless the flow of the program is interrupted or changed by a BASIC command in the program.

EXAMPLE 1.1

Consider the function $P = -x^2 + 12x + 64$. Let us write a BASIC program to evaluate this function for the whole numbers $0, 1, 2, \ldots, 10$.

```
5     REM ONE VARIABLE FUNCTION
6     REM EVALUATION AND PRINT OUT
10    FOR X=0 TO 10
20    P=-X**2+12*X+64
30    PRINT X,P
40    NEXT X
50    STOP
60    END
```

```
0               64
1               75
2               84
3               91
4               96
5               99
6               100
7               99
8               96
9               91
10              84
```

Lines 5 and 6 start with REM. In BASIC REM means remark or comment, and the computer types out but ignores whatever is typed on these lines. So the REM lines (which are optional) are used just for headings or labels.

Lines 10 and 40 form a FOR NEXT loop. FOR X=0 TO 10 tells the computer to perform lines 20, 30, and 40 eleven times with X=0 the first time, X=1 the second time, X=2 the third time, and so on until X=10 the eleventh time.

Line 20 is a translation of our algebraic equation $P = -x^2 + 12x + 64$ into BASIC. * is multiply, + is plus, – is minus, / is divide, and ** means raise to the power. Computation is done left to right on the right-hand side of the equal sign, except that ** is done first, then * or / and then + or –. Also, any expression inside parentheses is done before moving outside the parentheses. When the calculation on the right-hand side of the equal sign is completed, the number is assigned to the variable on the left-hand side of the equal sign. Line 30, PRINT X,P, prints the current values of X and P, each of the eleven times through the loop.

So the first time at line 10, X=0. Then at 20, $-x^2 + 12x + 64$ is evaluated with X=0. The resulting value of 64 is assigned to P. Then in line 30 the computer prints the current values of X and P, namely, 0 and 64, respectively. At NEXT X the computer goes back to line 10 and X becomes 1. Then at line 20, $-x^2 + 12x + 64$ is evaluated with X=1; the resulting value of 75 is assigned to P. Then in line 30 the computer prints the current values of X and P (1 and 75, respectively, in this case). Then X goes up to 2, P=84, and they are printed, and so on until X=10 and P=84. Then the FOR NEXT loop is satisfied, and the computer leaves the loop and goes to the next statement, which is STOP followed by END, and the program is done.

Now, of course, we could have evaluated this P function by hand for the eleven different X values without too much effort. However, if we changed line 10 to FOR X=0 to 100000, then the program would be ordered to evaluate and print P for X=0, 1, 2, 3, . . . , 100,000. This is just as easy for the computer, but it would take us days of computation time with pencil and paper. It is this tremendous speed and error-free calculation ability of the computer that we want to exploit.

Notice on the printout of the program that P is a maximum when X=6. Let us write a BASIC program to go through the eleven X values again and calculate P. This time, let us have it print just the optimum (maximum) solution.

```
5     REM ONE VARIABLE FUNCTION
6     REM TO FIND THE MAXIMUM
8     B=-999999
10    FOR X=0 TO 10
20    P=-X**2+12*X+64
30    IF P > B THEN 50
40    GO TO 70
50    A1=X
60    B=P
70    NEXT X
80    PRINT A1,B
90    STOP
100   END

6        100
```

Lines 5 and 6 label the program. Line 8 assigns B a very small number so that the first P value will be larger than B. Again we have a FOR NEXT loop in lines 10 through 70. With X=0, P=64 at line 30 IF P>B checks to see if the current value of P is greater than (> means greater than in BASIC) the current value of B. It is; therefore it goes directly to line 50 and stores X=0 in A1 and P=64 in B. Then NEXT X so X=1 at line 10, and $P = -x^2 + 12x + 64$ is evaluated for X=1 at line 20. Then at line 30, P=75 is greater than 64 so it goes to line 50 and erases A1=0 and B=64 and stores X=1 and P=75 in A1 and B, respectively. Then X=2 and P=84 are stored in A1 and B, respectively. Then X=3 and P=91 are stored in A1 and B, respectively. Then X=4 and P=96 are stored in A1 and B, respectively. Then X=5 and P=99 are stored in A1 and B, respectively. Then X=6 and P=100 are stored in A1 and B, respectively. Then X=7 and P=99, and in line 30, P=99 is not greater than the current value of B, namely, 100. Therefore, the logical expression P>B is false, so THEN 50 is ignored and the program just goes to the next line, line 40. There the command GO TO 70 (it is GOTO in some versions of BASIC) is followed; the computer skips to line 70 and X goes up to 8 and P=96, and P>B is false again so it skips 50 and 60 again. The same happens when X=9 and finally when X=10.

Then the FOR NEXT loop is satisfied so the computer goes to the next line (line 80) and prints the current values of A1 and B. But A1 and B are storing 6 and 100 (the optimum solution), so the optimum answer is all that the computer reports.

Again, the computer does this so quickly that we have a powerful new way of finding the maximum (or, similarly, the minimum) of functions.

If X=0, 1, 2, . . . , 100000, the computer can check those 100,001 values for the optimal answer and report back in a second or two. This frees us to state good questions and let the computer worry about the answer.

Notice that the program used the IF THEN BASIC statement. Any logical expression involving BASIC variables (A, B, C, . . . , Z, A1, B1, C1, . . . , Z1, A2, B2, C2, . . . , Z2, . . . , A9, B9, C9, . . . , Z9, A0, B0, C0, . . . , Z0) can be put between the IF and the THEN. If the logical expression is true, then the computer jumps to the line number stated after THEN. If the logical expression is false, then the computer goes to the statement immediately below the IF THEN. This gives you the power to branch all over the program at the appropriate time to accomplish your task. Also, the GO TO and then a line number is just what it says. As soon as the computer arrives at GO TO 70, for example, it goes directly to line 70, skipping everything else. Therefore, we can see that we only get to the storage area of the program (lines 50 and 60) when we have a P value which is greater than the maximum so far. This guarantees that the computer will find, store, and then print the true optimum solution.

EXAMPLE 1.2

A projectile is fired from 100 meters above the ocean floor in the Pacific. Its equation of motion is given by $P = -x^2 + 140x + 100$, where x is the time in seconds after firing and P is the height above the ocean floor in meters. How high will the projectile rise in the first 100 seconds and when will it reach its peak during that time? In symbols the problem is maximize $P = -x^2 + 140x + 100$ subject to $0 \leqslant x \leqslant 100$.

The BASIC program to solve this problem is given below:

```
5      REM ONE VARIABLE
6      REM MAXIMIZATION
10     B=-999999
20     FOR X=0 TO 100
30     P=-X**2+140*X+100
40     IF P > B THEN 60
50     GO TO 80
60     A1=X
70     B=P
80     NEXT X
90     PRINT A1,B
100    STOP
110    END

70          5000
```

Therefore, the projectile will reach a maximum height of 5000 meters above the ocean floor 70 seconds after launching. Notice that structurally we have the same problem as before. The computer just checks the P value for X=0, 1, 2, 3, . . . , 100 and stores the optimum in lines 60, A1=X, and 70, B=P, and then prints the optimum answer at the end of the loop. It does not matter where the optimum occurs (X=0, X=1, . . . , X=100); the computer will find it and store it, and then print it at the end of the program.

EXAMPLE 1.3

A patient is given medication and the doctor needs to know when the drug reaches its maximum concentration in the bloodstream. Several readings have been taken in the past and a polynomial equation fitted to describe concentration as a function of time. The concentration equation as a function of time in minutes is given by $P = -x^2 + 59.3x + 2558$.

It is known that the maximum concentration occurs sometime in the first 200 minutes after the patient takes the medicine. We are also interested in finding when this optimum occurs to the nearest tenth of a minute.

We have maximize $P = -x^2 + 59.3x + 2558$ subject to $0 \leqslant x \leqslant 200$. The program is given below:

```
5     REM ONE VARIABLE
6     REM MAXIMIZATION
7     REM BY INCREMENTS
8     REM OF ONE TENTH
10    B=-999999
20    FOR X=0 TO 200 STEP .1
30    P=-X**2+59.3*X+2558
40    IF P > B THEN 60
50    GO TO 80
60    A1=X
70    B=P
80    NEXT X
90    PRINT A1,B
100   STOP
110   END

29.7000        3437.12
```

Therefore, after 29.7 minutes the concentration reaches a maximum of 3437.12 ten thousandths of a percent.

Notice line 20, FOR X=0 TO 200 STEP.1. The STEP .1 is a feature of BASIC that allows X to be incremented by tenths (or any other increment specified) from 0 to .1 to .2, . . . , 1 to 1.1 to 1.2, . . . , 100. The rest of the program is the same as before.

EXAMPLE 1.4

Consider the problem to maximize $P = -5x^2 + 72x + 900$ subject to $0 \leqslant x \leqslant 300$. The program is given below:

```
5     REM ONE VARIABLE
6     REM MAXIMIZATION
10    B=-999999
20    FOR X=0 TO 300
30    P=-5*X**2+72*X+900
```

```
40   IF P  >  B THEN 60
50   GO TO 80
60   A1=X
70   B=P
80   NEXT X
90   PRINT A1,B
100  STOP
110  END

7          1159
```

Therefore, the maximum is $P = 1159$ at the point X=7.

EXAMPLE 1.5

Suppose that the Finland Company has studied their inventory costs for one product and feel that their costs can be described by $C = 3x^2 - 1117x + 200{,}000$, where x is the number of units of the product that are reordered each time an order is placed with their supplier. Their warehouse has room for only 614 units of the product at any one time. So we seek to minimize $C = 3x^2 - 1117x + 200{,}000$ subject to $0 \leqslant x \leqslant 614$. The program is below:

```
5    REM ONE VARIABLE
6    REM MINIMIZATION
10   B=999999
20   FOR X=0 TO 614
30   C=3*X**2-1117*X+200000
40   IF C < B THEN 60
50   GO TO 80
60   A1=X
70   B=C
80   NEXT X
90   PRINT A1,B
100  STOP
110  END

186        96026
```

Therefore, the company should place an order for 186 units of the product every time they reorder to keep their inventory costs down to a minimum of \$96,026 per year. This minimization program is the same in structure as our maximize programs except that we arrange to store the minimum solution in lines 60 and 70.

In line 10, B is assigned a very large number so that the first C value will be less than B and hence go to the storage area (60 and 70). Lines 20 through 80 set up a loop to go through X=0, X=1, X=2, . . . , X=614 to find the minimum C value. Notice in line 40, IF C<B THEN 60, that C<B (< means less than in BASIC) checks to see if the current C value is less than B=999999 or B equals the minimum so far after a few times through the loop.

Let us consider the function $P(x,y) = x^2 + 2x + xy - xy^2$ subject to $0 \leqslant x \leqslant 2$ and $0 \leqslant y \leqslant 2$. This is a function of two variables. The only possibilities for (X,Y) are (0,0), (0,1), (0,2), (1,0), (1,1), (1,2), (2,0), (2,1) and (2,2). Therefore, if we evaluated the function for these nine ordered pairs we could find the maximum. But we could write a BASIC program to maximize a two-variable function by just adding an extra FOR NEXT loop (one loop per variable). To illustrate this, let us try a bigger problem.

EXAMPLE 1.6

Maximize $P = 50 + \sqrt{997775 - x^2 + 80x - y^2 + 50y}$ subject to $0 \leqslant x \leqslant 500$ and $0 \leqslant y \leqslant 400$.

The possible ordered pairs are (0,0), (0,1), (0,2), . . . , (500,400). There are 200,901 ordered pairs in all. A BASIC program that looks at all 200,901 ordered pairs and finds the true optimum follows:

```
10    B=-999999
20    FOR X=0 TO 500
30    FOR Y=0 TO 400
40    P=SQR(997775-X**2+80*X-Y**2+50*Y)+50
50    IF P > B THEN 70
60    GO TO 100
70    B=P
80    A1=X
90    A2=Y
100   NEXT Y
110   NEXT X
120   PRINT A1,A2
130   PRINT B
140   STOP
150   END

40           25
1050
```

Therefore, $x = 40, y = 25$ yields the maximum P value of 1050.

Lines 20 through 110 form the outside loop that varies X from 0 to 1 to 2 to 500. Lines 30 through 100 form the inside loop that varies Y from 0 to 1 to 2 to 400 for each X value. In other words, the 30 through 100 inner loop is performed 501 x 401 = 200,901 times, each time with a different X, Y pair.

The rest of the program is the same as the previous ones. The storage space is 70, 80 and 90 this time.

EXAMPLE 1.7

Now let us maximize $P = 5x + 6y$ subject to $3x + y \leqslant 500$ and $x + 3y \leqslant 1000$. We can see that x must be less than 500/3. = 166.67. From $3x + y \leqslant 500$ we can see that y must be less than or equal to 500.

```
5     B=-999999
10    FOR X=0 TO 166
20    FOR Y=0 TO 500
30    IF X+3*Y > 1000 THEN 110
40    IF 3*X+Y > 500 THEN 110
50    P=5*X+6*Y
60    IF P > B THEN 80
70    GO TO 110
80    X1=X
90    Y1=Y
100   B=P
110   NEXT Y
120   NEXT X
130   PRINT X1,Y1,B
140   STOP
150   END

61          313              2183
```

Therefore, $x = 61, y = 313$ yields the maximum of $P = 2{,}183$. Notice that lines 30 and 40 have the two constraints written with their inequalities reversed. If X + 3*Y>1000 and 3*X+Y>500 are both false, then the program will get to line 50 and evaluate P. If, however, either one or both of those statements are true, then the computer will go to line 110, effectively skipping that particular ordered pair. This way only the ordered pairs that satisfy both $x + 3y \leqslant 1000$ and $3x + y \leqslant 500$ will be considered for the maximum.

EXAMPLE 1.8

Let us maximize $P = 600{,}000 - x^2 + 40x - y^2 + 60y - z^2 + 21z$ subject to $0 \leqslant x_i \leqslant 100$ for $i = 1, 2, 3$. Therefore, we will write a program to evaluate P for all 101 x 101 x 101 = 1,030,301 ordered triples (x,y,z) and find the optimum combination. The program follows:

```
10    B=-999999
20    FOR X=0 TO 100
30    FOR Y=0 TO 100
40    FOR Z=0 TO 100
50    P=600000-X**2+40*X-Y**2+60*Y-Z**2+21*Z
60    IF P > B THEN 80
70    GO TO 120
80    B=P
90    A1=X
100   A2=Y
110   A3=Z
120   NEXT Z
130   NEXT Y
140   NEXT X
150   PRINT A1,A2,A3
160   PRINT B
170   STOP
180   END

20          30          10
601410
```

Therefore, $x_1 = 20$, $x_2 = 30$, $x_3 = 10$ yields the maximum P value of 601,410. This time we have three nested FOR NEXT statements, one for each variable, to see that all the ordered triples (0,0,0) (0,0,1), (0,0,2), . . . , (100,100,100) are looked at. Notice that the inside 40 through 120 loop is executed 101 x 101 x 101 = 1,030,301 times.

As you can see, each program has the same structure, namely:

Initialization statements–usually just assign a very small value to the eventual maximum P value (we have been using B=–999999), or a very large number to the eventual minimum C value (we have been using B=999999), if we are minimizing instead of maximizing.

FOR NEXT loops–check all feasible solutions, one loop for each variable. Each FOR NEXT loop is to be completely nested inside the loop outside of it.

Constraints–these should be reversed in BASIC from what they are in algebra so that if any one is not satisfied the program jumps to the end of the loop. This way, if there is a series of constraints, only when all are satisfied will the possible (feasible) points get to the function evaluation stage.

The objective function (profit, cost, etc.) to be optimized.

Comparison statement (like IF P>B THEN 80)–checks to see if the current function value is the optimum so far.

Storage area–for the current optimum solution. Stores both the x coordinates and their resultant current optimum value.

Printing statements–a statement or statements to output the answer from the storage area.

Therefore, in every problem that has about 1,000,000 or less answers, we can write a BASIC program to go through all the answers and store and print the optimum solution. If the problem has many more than a million answers, we will use Monte Carlo optimization (explained in chapter 2) or multistage Monte Carlo integer programs (explained in chapter 12).

EXERCISES

1.1 Maximize $P = -7.1x^2 + 412x + 10{,}000$ subject to $0 \leq x \leq 200$.

1.2 Minimize $C = 14x^3 - 75x^2 + x + 70{,}000$ subject to $0 \leq x \leq 800$.

1.3 Maximize $P = -2x_1{}^2 + 104x_1 - 17x_2{}^2 + 100x_2 + 2x_1x_2 + 705$ subject to $0 \leq x_1 \leq 80$ and $0 \leq x_2 \leq 75$.

1.4 Maximize $P = 5x + 3y$ subject to $x + 2y \leq 100$ and $3x + y \leq 110$.

1.5 The West L.A. Retailers Group has developed the following price-quantity curve for Product 1: $P_1 = -.000078x_1 + 112$,* where x_1 is the number of units produced and sold and P_1 is the price (the idea here is that as one lowers the price, the number of units sold increases). West L.A. has a $9 unit cost and a fixed cost of $20,000 for Product 1. Find the maximum profit production solution and the corresponding price that they should charge.

1.6 The Trails End Ltd. people have developed the following price-quantity curve for their product: $P_1 = -.00029x_1 + 59$. Find the number of units to produce and sell (and the corresponding price) in order to maximize revenue.

1.7 White River Luxury Items has developed the following demand curve for their top luxury item: $P_1 = x^2 - 2000x + 1{,}000{,}000$ for $200 \leqslant x \leqslant 400$. Sales of this item are so sensitive to price that they would like to find the right price to maximize revenue. Write a BASIC program to search $x = 200, 201, \ldots, 400$ to find the maximum revenue solution.

1.8 Maximize $P = 2x + 2.2y$ subject to $x \geqslant 0$, $y \geqslant 0$, $x + .8y \leqslant 100$, and $2x + 3y \leqslant 350$.

1.9 Minimize $C = 7x + 3y$ subject to $x \geqslant 0$, $y \geqslant 0$, and $-2x^2 + 48x - 3y^2 + 75y + 100 \geqslant 1000$.

* See Appendix A.

Suggested Reading

Barnett, Eugene. *Programming Time-Shared Computers in BASIC*. New York: Wiley-InterScience, 1972.

Kapur, J. "Optimization in Mathematics, Statistics, Economics, Operations Research, Science and Technology." *International Journal of Mathematical Education in Science and Technology* 10 (1979): pp. 441-53.

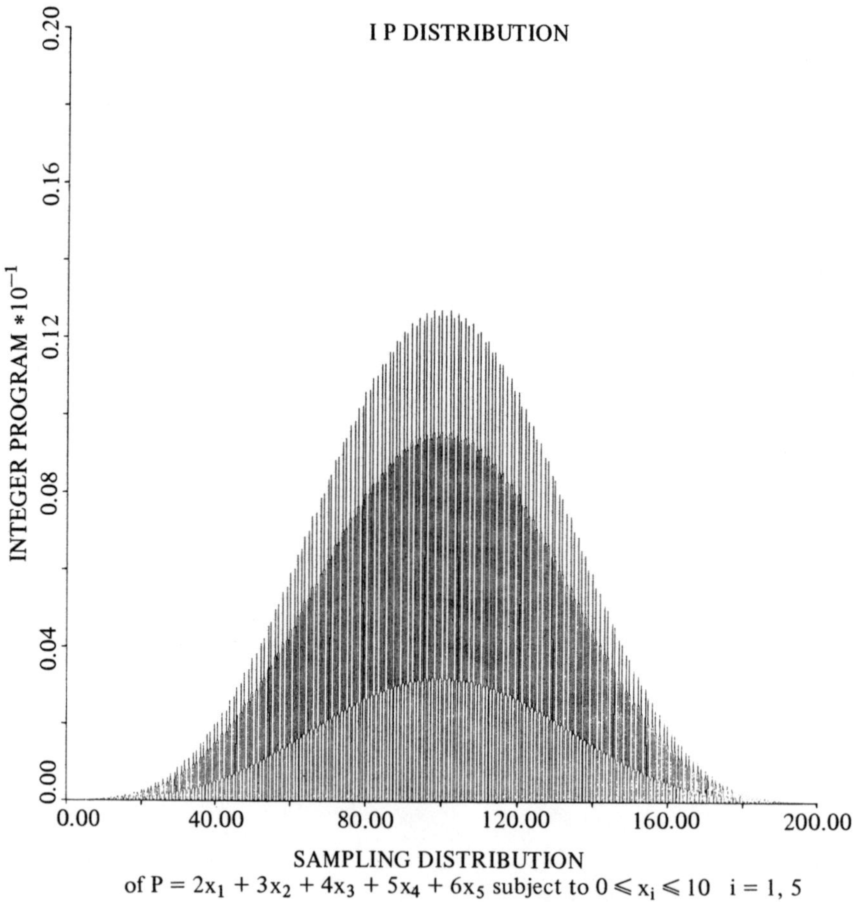

FIGURE 2.1 is a graph of the frequency of occurrence of all the feasible solutions of another integer program (all solutions must take whole number answers only). Notice that all the answers are connected (no isolated maximums or minimums). The fact that all the answers are connected (close together) will help us find nearly optimal solutions using Monte Carlo simulation techniques. It will also help us find the true optimal solution of large-scale problems in chapter 12.

CHAPTER 2

Monte Carlo Optimization

We begin this chapter with an example to review our "search all solutions" technique. Then we shall concentrate on the Monte Carlo technique to maximize or minimize functions. Along the way we shall review key BASIC commands and look at some examples.

EXAMPLE 2.1

Let us maximize $P = 55x_1^2 + 100x_1x_2 + 75x_2^2 + 195x_1 + 4x_2 + 18$ subject to $x_1 \geqslant 0$, $x_2 \geqslant 0$, $5x_1 + 2x_2 \leqslant 1000$, and $2x_1 + 5x_2 \leqslant 1000$.

```
1      REM MAXIMIZE A FUNCTION
2      REM SUBJECT TO CONSTRAINTS
10     B=-999999
20     FOR X1=0 TO 200
30     FOR X2=0 TO 200
40     IF 5*X1+2*X2 > 1000 THEN 140
50     IF 2*X1+5*X2 > 1000 THEN 140
60     P1=55*X1**2+100*X1*X2+75*X2**2
70     P2=195*X1+4*X2+18
80     P=P1+P2
90     IF P > B THEN 110
100    GO TO 140
110    B=P
120    A1=X1
130    A2=X2
140    NEXT X2
```

```
150  NEXT X1
160  PRINT A1,A2
170  PRINT B
180  STOP
190  END

142               143
4701575
```

As can be seen from $5x_1 + 2x_2 \leqslant 1000$, x_1 cannot exceed 200 = 1000/5. From $2x_1 + 5x_2 \leqslant 1000$, x_2 is also bounded by 200. Each loop then goes from 0 through 200 for the x values. Therefore, 201 x 201 = 40,401 ordered pairs are looked at, and lines 40 and 50 throw out the pairs that do not satisfy both $5x_1 + 2x_2 \leqslant 1000$ and $2x_1 + 5x_2 \leqslant 1000$. For ease of typing, the P function is split and then added in P=P1+P2. Lines 110, 120, and 130 are the storage space for the optimum so far. As before, a solution P gets to the storage space (with its attendant x_1 and x_2 values) only if P is greater than B (B is the current optimum). As with earlier programs, at each point the storage area has the best answer so far. Therefore, at the end of the run through all the possible solutions, the optimum so far is the true optimal solution.

Let us solve the above problem again, but this time let us try to get x_1 and x_2 to the nearest tenth. We have 2001 x 2001 = 4,004,001 answers to check. The program follows:

```
1     REM MAXIMIZE A FUNCTION
2     REM SUBJECT TO CONSTRAINTS
10    B=-999999
20    FOR X1=0 TO 200 STEP .1
30    FOR X2=0 TO 200 STEP .1
40    IF 5*X1+2*X2 > 1000 THEN 140
50    IF 2*X1+5*X2 > 1000 THEN 140
60    P1=55*X1**2+100*X1*X2+75*X2**2
70    P2=195*X1+4*X2+18
80    P=P1+P2
90    IF P > B THEN 110
100   GO TO 140
110   B=P
120   A1=X1
130   A2=X2
140   NEXT X2
150   NEXT X1
```

```
160  PRINT A1,A2
170  PRINT B
180  STOP
190  END

142.700          142.900
4.71911E+06
```

Everything is the same as before except that STEP .1 has been inserted in lines 20 and 30. Also, we could make it more accurate by rerunning the program with, for example, the following:

```
20 FOR X1=140 TO 145 STEP .01
30 FOR X2=140 TO 145 STEP .01
```

EXAMPLE 2.2

Consider the program to maximize $P = 4x_1 + 5x_2$ subject to $2x_1 + 6x_2 \leqslant 900$, $4x_1 + 3x_2 \leqslant 1000$, $0 \leqslant x_1 \leqslant 100$, and $0 \leqslant x_2 \leqslant 200$. The program is:

```
5       B=-99999
10      FOR X1=0 TO 100
20      FOR X2=0 TO 200
30      IF 2*X1+6*X2 > 900 THEN 170
40      IF 4*X1+3*X2 > 1000 THEN 170
50      P=4*X1+5*X2
60      IF P > B THEN 80
70      GO TO 170
80      A1=X1
90      A2=X2
100     B=P
170     NEXT X2
180     NEXT X1
200     PRINT A1,A2
210     PRINT B
220     STOP
230     END


99              117
981
```

Therefore, $x_1 = 99$ and $x_2 = 117$, with a corresponding $P = 981$ proved to be maximal.

Monte Carlo Approach to Optimization

The major problem with the technique we are using is that it is one of those "how high is up" arguments. Regardless of how fast a computer is or becomes in the future, there will always be problems that have so many answers that it would take a computer years to look at them all.

We propose to get around this problem through a technique called Monte Carlo Optimization (and with various refinements of that technique). In its simplest form it involves randomly looking for the optimal solution (we shall write BASIC programs to do this). In its most sophisticated form—the multistage Monte Carlo integer program (chapter 12)—one can find the *true optimum* solution to a problem with 1,000,000, 000,000,000,000,000,000,000,000)^2 feasible solutions in about a minute on a medium-sized computer! A search-all technique on a problem of this scale would take the computer about 32 million trillion years!

EXAMPLE 2.3

Consider the example to maximize $P = 2x_1 + 10x_2 + 25x_3 + 50x_4$ subject to $0 \leqslant x_i \leqslant 100$ for $i = 1, 2, 3, 4$. Now there are 101 x 101 x 101 x 101 = 104,060,401 feasible solutions to this problem, namely, (0,0,0,0), (0,0,0,1), (0,0,0,2), . . . , (100,100,100,100). In this case, the maximum is when $x_1 = 100$, $x_2 = 100$, $x_3 = 100$, $x_4 = 100$ with a corresponding $P = 8700$. This can be seen because there are only plus signs in the P function. However, let us pretend we did not know that and try to write a BASIC program for the answer.

In BASIC, if you set $x = 1$ in an early statement, then every time you type in RND (X) the computer supplies a random number between 0 and 1, for example, .2562147. Each number is equally likely to occur on each pass through an RND (X) statement. Also, in BASIC there exists the built-in function INT ( ). This statement drops the decimal part of any number. Therefore, a statement like INT (RND (X)*1000) would produce 256 from a random number like .2562147. The RND (X)*1000 would move the decimal over three places, 256.2147. Then the INT part would remove the .2147, leaving 256. The important point is that the RND (X)*1000 combined with the INT in this fashion would, each time that statement is executed, produce the numbers 0, 1, 2, 3, . . . , or 999 completely at random. Each of the 1000 numbers would be equally likely to occur on each execution of the statement. In general, you multiply by one more than the upper bound. So, if you want numbers 0, 1, 2, . . . , 999, multiply by 1000. If you want 0, 1, 2, 3, 4, 5, multiply by 6, etc. Now, getting back to our question, let us arrange to read in four random numbers from 0, 1, 2, . . . , to 100 (one for each of the variables) and evaluate *P* with them. Have the program do this tens of thousands of times and store and print the best optimum from this "random sample" of the feasible solutions. How close do we come to the answer? The program and printout from five separate runs of 10,000 each are shown below:

```
2      X=1
5      B=-999999
10     FOR I=1 TO 10000
20     X1=INT(RND(X)*101)
30     X2=INT(RND(X)*101)
40     X3=INT(RND(X)*101)
50     X4=INT(RND(X)*101)
60     P=2*X1+10*X2+25*X3+50*X4
70     IF P > B THEN 90
80     GO TO 140
90     AL=X1
100    A2=X2
110    A3=X3
120    A4=X4
130    B=P
140    NEXT I
150    PRINT A1,A2,A3,A4
160    PRINT B
170    STOP
180    END
```

```
59          88          99          100
8473

67          100         100         100
8634

99          98          96          99
8528

66          100         99          100
8607

59          89          100         100
8508
```

Line 2 sets x=1 to start the random number generator. Line 5 initializes B=-999999 to be a very small number to get the maximization going in the right direction.

In the Monte Carlo optimizations, we have only one loop (regardless of the number of variables). But we have a separate RND statement for each variable so that the four variable values are independent of each other each time. Lines 10 through 140 comprise this loop. We have selected to run this loop 10,000 times for convenience. 10,000 runs takes only a second or two on our computer, but it could be a higher or lower loop value as dictated by the problem and/or one's computer. Note that the I in FOR I=1 TO 10000 merely keeps track of how many times the computer has gone through the loop. It does not figure in the calculation as in previous problems.

20 X1=INT(RND (X)*101) yields either 0, 1, 2, 3, . . . , 100 completely at random and assigns it to X1.

30 X2=INT(RND (X)*101) yields either 0, 1, 2, 3, . . . , 100 completely at random and (independent from the X1 value) assigns it to X2.

Lines 40 and 50 accomplish the same for X3 and X4, respectively. All four numbers are completely independent of each other on each run. They may be the same once in a while, but it is purely by chance.

As you can see, this is a seemingly very inefficient way of trying to maximize a function. But the computer is so fast (and getting faster) that it really does a pretty good job of it.

Line 60 evaluates P for our random answer. Line 70 checks to see if it is the optimum solution so far. If it is, THEN 90 sends it to the storage area (90, 100, 110, 120, and 130) to store the four coordinates and the resulting P value. If not, GO TO 140 is followed and on to the next random solution.

The best of the five runs was the second run with P=8634. (Notice we are fairly close to the optimum of 8700 already.) The printout of variable four was, 100, 100, 99, 100, 100. So let us set X4=100 and run it again (thereby narrowing or focusing the Monte Carlo simulation run a bit). The program, with the only change being 50 X4=100, is below:

```
2     X=1
5     B=-999999
10    FOR I=1 TO 10000
20    X1=INT(RND(X)*101)
30    X2=INT(RND(X)*101)
```

```
40    X3=INT(RND(X)*101)
50    X4=100
60    P=2*X1+10*X2+25*X3+50*X4
70    IF P>B THEN 90
80    GO TO 140
90    A1=X1
100   A2=X2
110   A3=X3
120   A4=X4
130   B=P
140   NEXT I
150   PRINT A1,A2,A3,A4
160   PRINT B
170   STOP
180   END
```

```
99                98               100              100
8678

100               100              99               100
8675

100               99               100              100
8690

86                96               100              100
8632

87                97               100              100
8644
```

Now we are doing much better, as evidenced by the higher P values. Run 3 produced P=8690. Since x_3 looks pretty stable, let us insert 40 X3=100, focusing in even further:

```
2     X=1
5     B=-999999
10    FOR I=1 TO 10000
20    X1=INT(RND(X)*101)
30    X2=INT(RND(X)*101)
40    X3=100
50    X4=100
60    P=2*X1+10*X2+25*X3+50*X4
70    IF P>B THEN 90
80    GO TO 140
```

```
90   A1=X1
100  A2=X2
110  A3=X3
120  A4=X4
130  B=P
140  NEXT I
150  PRINT A1,A2,A3,A4
160  PRINT B
170  STOP
180  END

98            100           100           100
8696

100           100           100           100
8700

100           100           100           100
8700

100           100           100           100
8700
```

We get the true optimum this time. And this was all done in a matter of a minute or less, once the program was written and typed.

But is this a legitimate way to solve problems? Yes. Now, of course, if there had been cross-products ($x_1 \times x_2$), pinning down one variable might conceivably hurt our search for the true combination. However, the author's considerable investigation of the statistics of the situation (more about this later) indicates that that problem is not very likely. The other incredible fact is that this solution was produced in a minute. So if one is at all skeptical, just run the problem several more times. We shall deal with the possible problems with this technique later. Right now, let us try another problem.

EXAMPLE 2.4

Maximize $P = 14x_1 + 37x_2 + 81x_3 + 55x_4 + 40x_5$ subject to $0 \leq x_i \leq 60$. Again, the optimal answer is obviously $x_1 = 60, x_2 = 60, x_3 = 60, x_4 = 60, x_5 = 60$, with a corresponding $P = 13{,}620$. But let us try a Monte Carlo run for the optimum and see how close we can get to the right answer.

```
5    REM MONTE CARLO INTEGER PROGRAM
10   B=-999999
20   X=1
```

```
25    FOR I=1 to 100000
30    X1=INT(RND(X)*61)
40    X2=INT(RND(X)*61)
50    X3=INT(RND(X)*61)
60    X4=INT(RND(X)*61)
70    X5=INT(RND(X)*61)
80    P=14*X1+37*X2+81*X3+55*X4+40*X5
90    IF P > B THEN 110
100   GO TO 170
110   A1=X1
120   A2=X2
130   A3=X3
140   A4=X4
150   A5=X5
160   B=P
170   NEXT I
180   PRINT A1,A2,A3,A4,A5
190   PRINT B
200   STOP
210   END
```

```
60          60          60          60          57
13500

56          53          60          59          59
13210

60          60          59          57          57
13254

47          60          60          59          55
13183

47          60          60          59          57
13263

33          60          60          59          57
13067

53          60          59          58          60
13331

60          60          60          59          57
13445

54          60          60          59          60
13481

60          60          60          60          58
13540
```

The printout displays the results of 10 runs of 100,000 answers each time. The best answer was $x_1 = 60, x_2 = 60, x_3 = 60, x_4 = 60, x_5 = 58$, with $P = 13{,}540$. Notice that this is 99.41% of the true optimum answer. And, of course, we could find the optimum by focusing our search in the upper end of the solution space.

EXAMPLE 2.5

Now let us try a problem where the answer is not so obvious. Maximize $P = x_1{}^2 + x_1\ x_2 - x_2{}^2 + x_3x_1 - x_3{}^2 + 8x_4{}^2 - 17x_5{}^2 + 6x_6{}^3 + x_6x_5x_4$ $x_7 + x_8{}^3 + x_9{}^4 - x_{10}{}^5 - x_{10}x_5 + 18x_3x_7x_6$ subject to $0 \leqslant x_i \leqslant 99$ for $i = 1, 2, \ldots, 10$. Again we will use a loop of 100,000 for convenience. This time, however, we will arrange for the program to read in 10 random numbers (between 0 and 99) independently by use of INT(RND (X)* 100) as before. The program follows:

```
5     B=-999999
10    X=1
20    FOR I=1 TO 100000
30    X1=INT(RND(X)*100)
40    X2=INT(RND(X)*100)
50    X3=INT(RND(X)*100)
60    X4=INT(RND(X)*100)
70    X5=INT(RND(X)*100)
80    X6=INT(RND(X)*100)
90    X7=INT(RND(X)*100)
100   X8=INT(RND(X)*100)
110   X9=INT(RND(X)*100)
120   X0=INT(RND(X)*100)
130   P1=X1**2+X1*X2-X2**2+X3*X1
140   P2=-X3**2+8*X4**2-17*X5**2
150   P3=6*X6**3+X6*X5*X4*X7+X8**3
160   P4=X9**4-X0**5-X0*X5+18*X3*X7*X6
170   P=P1+P2+P3+P4
180   IF P > B THEN 200
190   GO TO 310
200   B=P
210   A1=X1
220   A2=X2
230   A3=X3
240   A4=X4
250   A5=X5
260   A6=X6
270   A7=X7
280   A8=X8
```

```
290  A9=X9
300  A0=X0
310  NEXT I
320  PRINT A1,A2,A3,A4,A5
330  PRINT A6,A7,A8,A9,A0
340  PRINT B
350  STOP
360  END
```

```
35          85          92          88          99
96          80          21          98          0
177067799

70          66          66          98          97
95          95          9           99          9
197580315

80          98          63          93          90
99          80          86          97          5
170192435

58          13          51          92          88
99          94          74          96          5
174982763

54          57          52          92          89
98          90          51          97          19
172240878

32          98          96          92          87
96          88          66          99          1
183798812

31          7           58          87          97
95          98          30          96          0
178292798

7           11          96          77          98
95          88          65          99          7
178867248

71          22          96          75          79
99          80          87          99          8
163060880

70          47          53          96          92
88          94          74          95          2
166830979
```

After these several runs and looking at the function, it appeared that $x_1 = 99, x_4 = 99, x_6 = 99, x_7 = 99, x_8 = 99$, and $x_9 = 99$ in the optimum solution. So the program was modified to include these facts and rerun as below. (Note that these changes could be made one at a time, with more care, to make sure the optimum is not missed. The author is lumping the changes together here to save space.)

```
5     B=-999999
10    X=1
20    FOR I=0 to 100000
30    X1=99
40    X2=INT(RND(X)*100)
50    X3=INT(RND(X)*100)
60    X4=99
70    X5=INT(RND(X)*100)
80    X6=99
90    X7=99
100   X8=99
110   X9=99
120   X0=INT(RND(X)*100)
130   P1=X1**2+X1*X2-X2**2+X3*X1
140   P2=-X3**2+8*X4**2-17*X5**2
150   P3=6*X6**3+X6*X5*X4*X7+X8**3
160   P4=X9**4-X0**5-X0*X5+18*X3*X7*X6
170   P=P1+P2+P3+P4
180   IF P > B THEN 200
190   GO TO 310
200   B=P
210   A1=X1
220   A2=X2
230   A3=X3
240   A4=X4
250   A5=X5
260   A6=X6
270   A7=X7
280   A8=X8
290   A9=X9
300   A0=X0
310   NEXT I
320   PRINT A1,A2,A3,A4,A5
330   PRINT A6,A7,A8,A9,A0
340   PRINT B
350   STOP
360   END
```

```
99          33          99          99          99
99          99          99          99          2
216300217

99          33          99          99          99
99          99          99          99          1
216300347
```

Notice that the P value has gone up from around 197,000,000 to 216,000,000. Also notice that $x_3 = 99$, $x_5 = 99$, and $x_{10} = 0$ would be good additional changes to make. These changes are reflected in the next runs shown below:

```
5     B=-999999
10    X=1
20    FOR I=1 TO 1000
30    X1=99
40    X2=INT(RND(X)*100)
50    X3=99
60    X4=99
70    X5=99
80    X6=99
90    X7=99
100   X8=99
110   X9=99
120   X0=0
130   P1=X1**2+X1*X2-X2**2+X3*X1
140   P2=-X3**2+8*X4**2-17*X5**2
150   P3=6*X6**3+X6*X5*X4*X7+X8**3
160   P4=X9**4-X0**5-X0*X5+18*X3*X7*X6
170   P=P1+P2+P3+P4
180   IF P > B THEN 200
190   GO TO 310
200   B=P
210   A1=X1
230   A2=X2
240   A3=X3
250   A4=X4
260   A5=X5
270   A6=X6
280   A7=X7
290   A8=X8
300   A9=X9
```

```
310  A0=X0
320  PRINT A1,A2,A3,A4,A5
330  PRINT A6,A7,A8,A9,A0
340  PRINT B
350  STOP
360  END

99          49          99          99          99
99          99          99          99          0
216300719

99          49          99          99          99
99          99          99          99          0
216300719
```

These two printouts represent the true maximal integer solution. A further modified run would indicate that *P* is optimal when $x_2 = 49.5$ (if one allows fractional answers).

These multiple-run Monte Carlo integer programs were presented to demonstrate the power and the usefulness of the random search for optimal technique.

Even though some of these answers were obvious before we started the runs (again, that was for demonstration purposes), this process works just as well on problems where the optimum is completely unknown.

In fact, in chapter 12 we automate this process completely, and the computer does multiple runs and focuses on the true but unknown optimum solution in a matter of minutes or seconds. All this takes place without the programmer making any focusing decisions. It is all automated.

In chapter 12 these automated multistage integer programs find the true optimal solution of five- and ten-variable nonlinear functions with 1×10^{30} feasible solutions. The optimum is found in a few minutes (at most) each time. With Monte Carlo techniques and more sophisticated automated multistage integer programs, it is possible to solve any nonlinear optimization problem of up to 10, 15, 20 or 100 variables using only a few minutes of computer time. This frees the quantitative worker or model builder to use calculus and/or linear programming, when appropriate, and search-all Monte Carlo or multistage Monte Carlo techniques on the tough optimization problems.

For the first time in mathematics, we are free to ask the right question and let the computer do the work solving it. By doing this, it is possible to solve problems in chemical modeling, inventory analysis, capital budgeting, medical research, econometrics, and physics, to name just a few, that were never before possible.

Before moving on, let us go over two short examples and review the rules of BASIC.

EXAMPLE 2.6

The Appensell Company has analyzed its inventory costs for Item C and learned that each reorder costs $55. Also, the holding cost is $800 per unit per year, with Q/2 being the average number of units in the warehouse at any time. The demand for C in one year will be approximately 10,000 units. Therefore, how many units of Item C should be ordered at each reorder time to minimize inventory cost?

We have to minimize $C = 800Q/2 + 55 \times 10{,}000/Q$ for $1 \leqslant Q \leqslant 10{,}000$, where $800Q/2$ is the annual holding cost and $55 \times 10{,}000/Q$ is the annual ordering cost. The program is below:

```
5     REM STANDARD INVENTORY COST
6     REM MINIMIZATION WITH Q
7     REM THE NUMBER OF UNITS
8     REM IN EACH ORDER
10    B=999999
20    FOR   Q=1 TO 10000
30    C=800*Q/2+55*10000/Q
40    IF C < B THEN 60
50    GO TO 80
60    Q1=Q
70    B=C
80    NEXT Q
90    PRINT Q1
100   PRINT B
110   STOP
120   END

37
29664.9
```

Therefore, the company should order $Q = 37$ units each time it reorders in order to obtain a minimum cost of $C = \$29{,}664.90$ per year.

Lines 5, 6, 7, 8 use REM to label the program. The computer prints, but ignores these lines.

Line 10 sets B = 999999, a very large number, so that in line 40, C will be less than B and THEN follow 60 to store the first solution (as possible optimum) in Q1 and B. That solution will stay in Q1 (for Q) and B (for C) until a better one (less cost) comes along.

Lines 20 and 80 set up a FOR NEXT loop that varies Q from 1 to 2 to 3 to 4 . . . and so on until Q=10,000. And for each different Q value the program executes 20, 30, 40, 50, 60, 70, and 80 (of course, skipping lines as ordered to by the logic of the program). Each time, line 30 evaluates C for a different value of Q. Then line 40 checks to see if this current C value is smaller than B (the smallest C value so far). If it is, then the program jumps to line 60 and replaces the old best minimum so far with the new one. Then 80 and next Q back to line 20 and Q is incremented by one, and the process starts over again.

If, however, in line 40 C is not less than B, then the IF statement is false, so it ignores THEN 60 and merely goes to the next line (line 50) which says GO TO 80 and jumps over the storage space.

So a Q value and its corresponding C value only gets to the storage space (lines 60 and 70) if it is better (C<B) than the minimal cost so far.

This process is done for $Q = 1, 2, 3, 4, \ldots, 10{,}000$ so that at the end of the program lines 90 and 100 print the optimum solution so far, which at that point is the optimum solution because all solutions have been checked. Lines 110 and 120 merely end the program.

EXAMPLE 2.7

Consider the program to maximize $P = 16x_1 + 2x_2 + 9x_3$ subject to all $x_i \geqslant 0$ and $3x_1 + x_2 + 2x_3 \leqslant 100$. From $3x_1 + x_2 + 2x_3 \leqslant 100$ we can see that x_1, x_2, and x_3 can be no larger than 33 from (100/3), 100 from (100/1), and 50 from (100/2), respectively.

The program (with three nested loops—one for each variable) is below:

```
10    B=-999999
20    FOR X1=0 TO 33
30    FOR X2=0 TO 100
40    FOR X3=0 TO 50
50    IF 3*X1+X2+2*X3 > 100 THEN 130
60    P=16*X1+2*X2+9*X3
70    IF P > B THEN 90
80    GO TO 130
90    B=P
```

```
100  A1=X1
110  A2=X2
120  A3=X3
130  NEXT X3
140  NEXT X2
150  NEXT X1
160  PRINT A1,A2,A3
170  PRINT B
180  STOP
190  END

32             0            2
530
```

Line 10 sets B=–999999, a very small number, so that the first P value will be larger than B in line 70 and hence go to line 90 and store P in B and the x_1, x_2, x_3 coordinates in A1, A2, A3, respectively. This will be the optimal solution until a better one comes along.

Lines 20 through 150 set up the outer loop which is executed 34 times (once with $x_1 = 0$, once with $x_1 = 1, \ldots$, and once with $x_1 = 33$). Lines 30 through 140 set up the middle inner loop which is executed 34 x 101 times (with x_2 varying from 0 to 1 to 2 to . . . 100 for each of the 34 times that $x_1 = 1, x_1 = 2, \ldots, x_1 = 33$ in the outer loop). Lines 40 through 130 set up the innermost loop which is executed 34 x 101 x 51 times (with x_3 varying from 0 to 1 to 2 to 50 for each of the 34 x 101 combinations of values for x_1 and x_2). So line 50 is run 34 x 101 x 51 times: the first time with $x_1 = 0, x_2 = 0, x_3 = 0$; then with $x_1 = 0, x_2 = 0, x_3 = 1$; then with $x_1 = 0, x_2 = 0, x_3 = 2$, and so on until $x_1 = 0, x_2 = 0, x_3 = 50$; then with $x_1 = 0, x_2 = 1, x_3 = 0$; then $x_1 = 0, x_2 = 1, x_3 = 1$; then with $x_1 = 0, x_2 = 1, x_3 = 2$, and so on until $x_1 = 0, x_2 = 1, x_3 = 50$; then with $x_1 = 0, x_2 = 2, x_3 = 0$, and so on until all 34 x 101 x 51 triples (ending with $x_1 = 33, x_2 = 100$, and $x_3 = 50$) have run through the IF statement in line 50.

Each time through line 50, if $3x_1 + x_2 + 2x_3$ is greater than 100 then the computer jumps to line 130 for NEXT X3 and the computer goes on to the next ordered triple (x_1, x_2, x_3) because this triple does not satisfy the constraint $3x_1 + x_2 + 2x_3 \leqslant 100$. However, if $3x_1 + x_2 + 2x_3$ is less than or equal to 100, then the IF (line 50) is false, so THEN 130 is ignored and the computer goes to the next line, 60, and evaluates the profit function $P = 16x_1 + 2x_2 + 9x_3$. Then line 70 checks to see if this current value of P is greater than the best maximum so far, namely,

B. If it is, then the computer jumps to line 90 and stores this new best optimal solution in B, A1, A2, and A3 in lines 90, 100, 110, and 120, respectively. However, if P is not greater than B then the IF (line 70) is false, so THEN 90 is ignored and the computer goes to the next line, 80. Line 80 has the computer GO TO 130, effectively skipping the storage space. Again, all combinations are checked and only the best answers (maximum in this case) so far are stored in the storage space (lines 90, 100, 110, and 120).

When lines 160 and 170 print the contents of the storage space at the end of the program, the computer is printing the optimum of all possible answers. Therefore, by definition, it is the maximum of this function under these constraints.

In this case $x_1 = 32$, $x_2 = 0$, $x_3 = 2$ yielded the maximum P value of 530.

Review of Key BASIC Commands

BASIC is a language that is useful for function evaluation and data manipulation. Its chief utility derives from its simplicity and yet great power because of the tremendous speeds at which a computer will execute BASIC commands. Frequently, speeds are in excess of one million operations per second.

BASIC statements are numbered in some ascending order and typed into the computer either at a terminal or on computer punch cards or some other manner. The commands are executed sequentially from top to bottom (one after another) unless the flow of the program is interrupted by a command which orders the computer to another area of the program. The three major control commands are (1) the FOR NEXT command sequence, (2) the IF THEN statement, and (3) the GO TO some number command.

An example of the FOR NEXT control command would be

```
10 FOR A=0 TO 100
20
30
40
50 NEXT A
```

Here lines 10, 20, 30, 40, and 50 would be executed in order by the computer 101 different times. The first time A=0, the second time A=1, then A=2, and so on through A=100. Lines 20, 30, and 40 would, for example, contain any statements that were appropriate to solving the problem at hand.

An example of the IF THEN statement would be IF C>D THEN 95. If the current value of C were greater than the current value of D, then the computer would jump to line 95. If C>D were false, however, then the computer would go on to the next statement.

Below is a list of symbols used in IF THEN statements:

>	greater than
> = or = >	greater than or equals
<	less than
< = or = <	less than or equals
=	equals
+	plus
-	minus
/	divide
*	multiply
**	raise to a power (exponentiation)
( )	parentheses which can be used as in algebra (operations inside the parentheses are performed first)

Any combination of these symbols and BASIC variables can be used in the IF THEN statement. This gives the programmer two-way branching control or multiple-way control with more than one if statement.

The GO TO is an unrestricted GO TO. No condition is tested. For example, it is just GO TO 50, meaning jump to line 50. With some computer systems, the GO TO is written as GOTO.

Any letter, or any letter followed by a whole number 0, 1, 2, 3, 4, 5, 6, 7, 8 or 9, can be used as a BASIC variable, for function evaluation, or data manipulation. Examples of BASIC variables are:

X
A
B
C
D1
K2
F6

(Note: In some versions of BASIC, so-called Extended BASIC or BASIC Plus, longer strings of letters or numbers are allowed for variables).

PRINT is used to print the results of the computer program. For example, PRINT X,Y would print the current values of the BASIC variables X and Y.

READ is used to read in data from the DATA line. For example,

```
10 READ X
50 DATA 10, 20, 30, 9, 28, 2, 16
```

would assign 10 to X the first time READ X is encountered in the program. The second time, 20 would be assigned to X, and so on. The DATA line can be placed anywhere in the program, but usually near the end.

STOP and END are put at the conclusion of a program. In some systems both are not necessary.

REM is again just used for comments. The computer types out the REM line but ignores it actually.

RND (X) will produce a number between 0 and 1, like .7258931, at random each time the statement is encountered. The only stipulation is that X=1 be assigned somewhere earlier in the program. Each time, a 7-digit number (can be more digits) is generated, and each one has one chance in 10,000,000 of occurring each time RND (X) is encountered. So in our optimization programs, if we wanted a number between 0 and 99 inclusive, we would multiply RND (X) by 100. Always multiply by one more than the upper bound. In this case, RND (X)*100 would yield 72.58931. If you want only random whole numbers, then INT ( ) can be employed. This basic built-in function chops off any fractional part of a number, so INT (RND (X)*100) would yield 72 in this case. Each successive run through the loop containing INT (RND (X)*100) would yield a new random number either 0, 1, 2, 3, 4, . . . , 99.

A model is a mathematical equation with or without constraints (usually inequalities or equations). Our goal is to maximize (or minimize) the equation within the boundaries of the constraints. The equation to be optimized (find maximum or minimum) is usually a profit function, cost function, or some other type of yield function.

These functions are arrived at by careful quantitative analysis of the practical problem at hand. This can include statistical forecasting of part or all of the model, i.e., trying to fit the best curve to the data and testing it for validity as a model. (Appendix A provides a brief introduction to statistical forecasting.) The other models are usually developed by a person knowledgeable in the particular area at hand. Developing good models is an art. But we hope to demonstrate in this text that maximizing or minimizing them is a science.

EXERCISES

2.1 Maximize $P = 3x^2 + 2xy - 4y^2 + 1800$ subject to $x + 2y \leqslant 700{,}000$ and $x \geqslant 0$, $y \geqslant 0$.

2.2 Maximize $P = 82x^2 + 714x + 1000$ subject to $x \geqslant 0$ and $x \leqslant 9000$.

2.3 Minimize $C = x^2 - 160x + 8000$ subject to $x \geqslant 0$ and $x \leqslant 100$.

2.4 Maximize $P = 6x_1 + 5x_2 + 7x_3$ subject to $x_i \geqslant 0, i = 1, 2, 3$, and $x_1 + 1.3x_2 + 1.6x_3 \leqslant 75{,}000$, and $2x_1 + 3x_2 + 5x_3 \leqslant 250{,}000$.

2.5 Minimize $C = 7x + 9y$ subject to $x \geqslant 0, y \geqslant 0$, $3x + 10y \leqslant 100$, and $10x + 2y \leqslant 99$.

2.6 Maximize $P = 10x_1 + 7x_2 + 9x_3 + 14x_4 + 17x_5 + 29x_6$ subject to all $x_i = 0$ or 1, and $50x_1 + 70x_2 + 100x_3 + 200x_4 + 250x_5 + 280x_6 \leqslant 500$.

2.7 Maximize $P = 2x_1 + 7x_2 + 14x_3 + 9x_4$ subject to $0 \leqslant x_i \leqslant 1{,}000{,}000$.

Hint. Do repeated Monte Carlo simulations to see how close to the true optimum, 32,000,000, you can get.

2.8 The Wakefield Company's manager estimates that if a price of \$30 is charged for their toaster, they will sell 1000 next year. The manager estimates that if a price of \$25 is charged, they will sell 10,000 toasters. The unit cost to make toasters is \$7. Find the maximum profit price and the resulting quantity that Wakefield should make.

Hint. Let us estimate the price-quantity curve for the toasters by fitting a straight line to the two points. $m = (y_2 - y_1)/(x_2 - x_1)$ is the slope of the line where (x_2, y_2) is (10,000,25) and (x_1, y_1) is (1000,30). Therefore, $m = (25 - 30)/(10{,}000 - 1000) = -.0005556$. From the straight line formula $y = mx + b$ (b is the y-intercept), we have $y = .0005556x + b$. Substituting in (10,000,25), we get $25 = -.0005556 \text{x} 10{,}000 + b$. So $b = 30.56$. Therefore, $y = -.0005556x + 30.56$ is the price-quantity curve. Therefore, $yx - 7x$ will be the profit function. We have maximize $P = -.0005556x^2 + 23.56x$ subject to $0 \leqslant x \leqslant 42{,}405$ (the point at which P drops to zero).

2.9 The president of Bessemer Northern Company estimated demand for product T as a function of price as follows:

Price (y)	Quantity demanded (x)
\$900	17,000
\$625	17,500
\$400	18,000
\$225	18,500
\$100	19,000

Using least squares nonlinear regression forecasting theory, the president fits the best quadratic equation to this data. This yields $y = .0001x^2 - 4x + 40{,}000$ for the price-quantity curve. Due to the limited range of demand (this is a luxury item), the president believes that x will be between 17,000 and 19,000. If the product costs the company \$200 a unit to produce, what is the optimal price to maximize profit? We have Profit $= xy - 200x = .0001x^3 - 4x^2 + 39{,}800x$ subject to $17{,}000 \leqslant x \leqslant 19{,}000$. Maximize this function.

2.10 The Brockway Mountain School cafeteria is planning to serve a meal consisting of meat, vegetable, and dessert. It has five different meat dishes to choose from, four vegetables, and three selections for dessert. However, the selections must contain at least a total of 100 units of Nutrient A, 200 units of Nutrient B, and 150 units of Nutrient C. The following chart lists each food selection with the amount of each nutrient it contains and also tells the price per serving:

	Nutrient A	Nutrient B	Nutrient C	Cost per serving
Meat 1	32	85	62	68¢
Meat 2	48	92	73	75¢
Meat 3	52	104	58	84¢
Meat 4	47	68	100	92¢
Meat 5	35	95	29	82¢
Vegetable 1	29	60	60	49¢
Vegetable 2	22	25	60	52¢
Vegetable 3	28	15	58	40¢
Vegetable 4	42	55	40	35¢
Dessert 1	22	48	55	46¢
Dessert 2	29	47	65	38¢
Dessert 3	50	40	38	28¢

Which combination of one meat item, one vegetable item, and one dessert item will fulfill the nutrient requirements and minimize the cost to the school cafeteria?

Hint. 12 variables, 12 nested loops, with x's = 0 or 1 only.

2.11 The Perry Sound company produces three products, A, B, and C. Product A yields a profit of \$30 per unit, while B and C yield profits of \$20 and \$15, respectively. The table below shows how many hours are required in each of two departments to make each unit of the three products:

Department	Product A	Product B	Product C
Department 1	7	7	3
Department 2	3	1	2

Department 1 has no more than 100 man-hours available per day. Department 2 has no more than 50 man-hours per day available. Find the number of A, B, and C that should be made to maximize the profit. The relevant information can be summarized as follows: Maximize $P = 30x + 20y + 15z$ subject to $x \geqslant 0$, $y \geqslant 0$, $z \geqslant 0$, $7x + 7y + 3z \leqslant 100$, and $3x + y + 2z \leqslant 50$. We have just written in equations and inequalities the information given in the story problem.

2.12 The Goegebic-Finland Company has a planning problem. Suppose that the price of Product A is related to the quantity sold in the sense that as the price is decreased the quantity sold increases. The daily price-quantity curve for Product A is $y = -.1x_1 + 100$, where x_1 is the quantity sold and y is the price.

Suppose also that the price of Product B is related to the quantity sold in the sense that as the price is increased the quantity sold increases (as with some fad items). The daily price-quantity curve for Product B is $y = .05x_2 + 110$.

However, products A and B are made in departments 1 and 2. There are 1000 minutes per hour available in Department 1 and 2000 minutes per hour available in Department 2. The time requirements for each product are shown below:

	Time required (minutes)	
Product	Department 1	Department 2
A	1	2
B	1	3

Product A costs \$.25 per unit to make and Product B costs \$.10 per unit to make. Therefore:

$$p_a = x_1y - .25x_1$$
$$p_a = x_1(-.1x_1 + 100) - .25x_1$$
$$p_a = -.1x_1{}^2 + 100x_1 - .25x_1$$
$$p_a = -.1x_1{}^2 + 99.75x_1 \quad \text{and}$$
$$p_b = x_2y - .10x_2$$
$$p_b = x_2(.05x_2 + 110) - .10x_2$$
$$p_b = .05x_2{}^2 + 110x_2 - .10x_2$$
$$p_b = .05x_2{}^2 + 109.9x_2$$

Therefore

$$p = p_a + b_b$$
$$p = -.1x_1{}^2 + 99.75x_1 + .05x_2{}^2 + 109.9x_2$$

We seek to maximize p subject to $x_1 \geqslant 0, x_2 \geqslant 0, x_1 + x_2 \leqslant 1000$, and $2x_1 + 3x_2 \leqslant 2000$.

Suggested Reading

Carter, L.R. and Huzan, E. *A Practical Approach to Computer Simulation in Business*. London: George Allen & Unwin Ltd., 1973.

Collatz, L. and Wetterling, W. *Optimization Problems*. New York: Springer-Verlag, 1975.

Kim, Chaiho. *Quantitative Analysis for Managerial Decisions*. Reading, Mass.: Addison-Wesley, 1976.

Eck, Roger. *Operations Research for Business*. Belmont, Calif.: Wadsworth, 1976.

Lancaster, Kelvin. *Modern Economics Principles and Policy*. New York: Rand McNally, 1973.

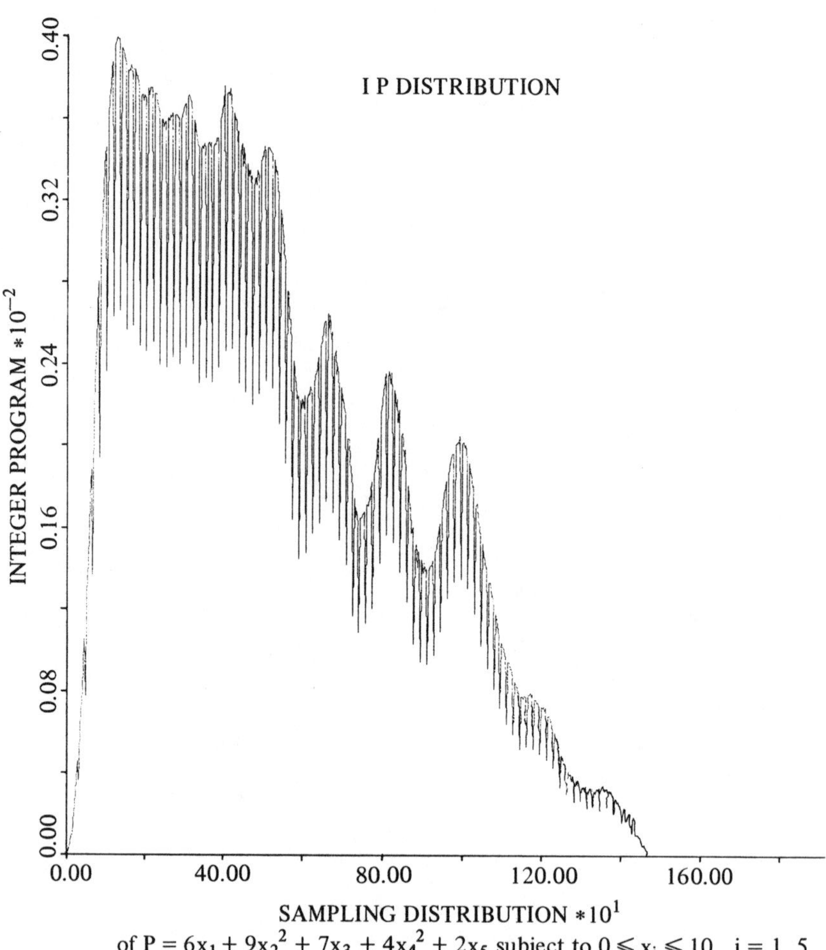

FIGURE 3.1 is the graph of the sampling distribution of $P = 6x_1 + 9x_2^2 + 7x_3 + 4x_4^2 + 2x_5$ subject to $0 \leqslant x_i \leqslant 10$, $i = 1{,}5$. Note that the nonlinear function makes the distribution (of all answers) asymmetric. But the important point is that the distribution is connected (no isolated optimums). (Reprinted by permission of the National Bureau of Standards, Special Publication 503, p. 364.)

CHAPTER 3

Business Applications and Ideas

We have seen optimization problems of the type to maximize $P = 6x + 4y$ subject to, say, $x \geqslant 0$, $y \geqslant 0$, $x + 5y \leqslant 10$, and $3x + 12y \leqslant 28$. These are problems with inequality constraints of some kind. However, there are problems which could have more rigid constraints, particularly in industrial or scientific research settings. As an example, the constraints might be $x + 5y = 10$ and $3x + 12y = 28$. Here strict equality is required. We can deal with these so-called equality constraints, either with or without attendant inequality constraints, in optimization problems. There are a number of ways to deal with them, depending on the problem. However, one of the more efficient ways to solve a problem with equation constraints in it is to solve the system of equations first and then do the problem. To that end we present a short discussion below. Then we shall return to optimization problems.

How to solve a system of equations

For a thorough treatment of solving systems of equations, the reader should consult one of the many fine books on linear algebra. Here a brief review of the "diagonal technique" is presented. There are, of course, other ways to solve systems of equations, but this technique has the additional benefit of being easy to program into a computer.

EXAMPLE 3.1

Let us consider an example. Solve the following system of equations:

$$x + 5y = 10$$
$$3x + 12y = 28$$

First we write down the coefficients (with all the x's in one column, the y's in another column, and equal to a constant column) without the variables, plus signs, or equal signs, which are all understood. This gives us:

$$\begin{matrix} 1 & 5 & 10 \\ 3 & 12 & 28 \end{matrix}$$

Our objective is to rearrange and/or change this matrix of numbers into the following:

$$\begin{matrix} 1 & 0 & a \\ 0 & 1 & b \end{matrix}$$

This will give the solution $x = a$ and $y = b$ if we use only the three elementary matrix operations which at any stage keep the solution space the same as it was with the beginning system of equations. The rules are as follows:

1. At any point, any two rows can be interchanged without changing the solution space.
2. At any point, any row can be multiplied or divided by any number (except zero) without changing the solution space.
3. At any point, any row can have added to it a multiple of another row without changing the solution space.

The idea is to put a 1 in the upper-left corner of the matrix and then 0's underneath it. Then put the next 1 down the main diagonal and then zeros everywhere else in that column and so on until you have something like this:

$$\begin{matrix} x & y & z & w & u & \\ 1 & 0 & 0 & 0 & 0 & 6 \\ 0 & 1 & 0 & 0 & 0 & 4 \\ 0 & 0 & 1 & 0 & 0 & 3 \\ 0 & 0 & 0 & 1 & 0 & 9 \\ 0 & 0 & 0 & 0 & 1 & -2 \end{matrix}$$

At this point the solution to this system would be:

$x = 6$
$y = 4$
$z = 3$
$w = 9$
$u = -2$

Now look back to the example of

$x + 5y = 10$

$3x + 12y = 28$

So

1	5	10
3	12	28

–3 times row 1 added to row 2
(Rule 3)

1	5	10
0	–3	–2

row two divided by – 3
(Rule 2)

1	5	10
0	1	2/3

–5 times row 2 added to row 1
(Rule 3)

1	0	20/3
0	1	2/3

Therefore, $x = 20/3$ and $y = 2/3$ is the solution to this system.

EXAMPLE 3.2
Solve the following system:

$2x + 14y = 30$

$2x + 9y \;\; = 60$

So

2	14	30
2	9	60

row 1 divided by 2
(Rule 2)

1	7	15
2	9	60

– 2 times row 1 added to row 2
(Rule 3)

1	7	15
0	-5	30

row 2 divided by – 5
(Rule 2)

1	7	15
0	1	-6

–7 times row 2 added to row 1
(Rule 3)

1	0	57
0	1	-6

Therefore, $x = 57$ and $y = -6$ is the solution to this system.

EXAMPLE 3.3
Solve the following system:

$$x + 3y + 5z = 100$$
$$2x + 9y + 30z = 50$$

When the number of equations does not equal the number of unknowns, just proceed as before (trying to put 1's down the diagonal and 0's everywhere else) and then interpret the results. Therefore, we have:

1	3	5	100
2	9	30	50

add –2 times row 1 to row 2
(Rule 3)

1 3 5 100
0 3 20 −150

divide row 2 by 3
(Rule 2)

1 3 5 100
0 1 20/3 −50

add −3 times row 2 to row 1
(Rule 3)

1 0 −15 250
0 1 20/3 −50

Therefore

$$x - 15z = 250$$

$$y + (20/3)z = -50$$

With z arbitrary

$$x = 15z + 250$$

$$y = -(20/3)z - 50$$

In this case, we have many solutions. By letting z take different values, the appropriate x and y values are produced from $x = 15z + 250$ and $y = -(20/3)z - 50$.

EXAMPLE 3.4

Suppose you started with four equations and six unknowns and the final matrix was as below:

x	y	z	w	u	v	
1	0	0	0	6	3	21
0	1	0	0	4	-2	27
0	0	1	0	8	3	-12
0	0	0	1	5	16	8

This would then represent:

$$x + 6u + 3v = 21$$

$$y + 4u - 2v = 27$$

$$z + 8u + 3v = -12$$

$$w + 5u + 16v = 8$$

So the solution would be for abritrary u and v:

$x = 21 - 6u - 3v$

$y = 27 - 4u + 2v$

$x = -12 - 8u - 3v$

$w = 8 - 5u - 16v$

EXAMPLE 3.5

Now let us try an optimization problem with some equality constraints.

Product A returns \$71 profit per unit, Product B returns \$52 profit per unit, and Product C yields \$29 per unit. Product A spends 10 minutes in Department 1 and 12 minutes in Department 2. Product B spends 15 minutes in Department 1 and 20 minutes in Department 2. Product C spends 20 minutes in Department 1 and 14 minutes in Department 2. There are 1500 worker-minutes available per time period in Department 1 and 1600 worker-minutes per time period in Department 2.

Due to a union contract and management policy, no workers are to be idle and none can be shifted from one department to another. Consequently, it is desired to find the product mix that maximizes profit and uses all of the 1500 + 1600 worker-minutes subject to the current worker deployment.

In addition, there are fixed costs of \$200, \$110, and \$80 per time period for production of products A, B, and C, respectively. These fixed costs do not have to be paid if no units of the particular product are made. The products are divisible and can be made in tenths of a unit.

The problem can be summarized as maximize $P = 71x + 52y + 29z - 200x_a - 110y_a - 80z_a$ subject to $x \geqslant 0$, $y \geqslant 0$, $z \geqslant 0$, and $x_a = 0$ if $x = 0$, $y_a = 0$ if $y = 0$, $z_a = 0$ if $z = 0$, $x_a = 1$ if $x > 0$, $y_a = 1$ if $y > 0$, $z_a = 1$ if $z > 0$, and $10x + 15y + 20z = 1500$ and $12x + 20y + 14z = 1600$.

Solving the system of equations we get:

60	90	120	9000
60	100	70	8000
60	90	120	9000
0	10	-50	-1000
60	0	570	18000
0	10	-50	-1000

x	y	z	
6	0	57	1800
0	1	–5	–100

Therefore

$$6x + 0y + 57z = 1800$$
$$0x + y - 5z = -100$$

For arbitrary z

$$x = \frac{1800 - 57z}{6}$$
$$y = 5z - 100$$

Therefore

$$20 \leqslant z \leqslant 31.58 \quad \text{then}$$
$$x = \frac{1800 - 57z}{6}$$
$$y = 5z - 100$$

The program to maximize P follows:

```
5     REM FIXED COST MAXIMIZATION
6     REM WITH EQUALITY CONSTRAINTS
8     B=-999999
10    FOR I=200 TO 316
30    Z=I/10.
40    X=(1800-57*Z)/6.
50    Y=5*Z-100
60    P=71*X+52*Y+29*Z-390
70    IF Y < .01 THEN 90
80    IF X < .01 THEN 110
85    GO TO 120
90    P=P+110
100   GO TO 120
110   P=P+200
120   IF P > B THEN 140
130   GO TO 175
140   A1=X
150   A2=Y
160   A3=Z
170   B=P
```

```
175   NEXT I
180   PRINT A1,A2,A3
190   PRINT B
200   STOP
210   END

110              0          20
8110
```

Therefore, the company should make $x = 110$ units of Product A and $z = 20$ units of Product C to maximize P at \$8110 per time period. Even though this is the optimal solution to the stated question, management is concerned that this might not be their most effective production strategy.

With that in mind, they have decided to rerun the computer program with the equation conditions $10x + 15y + 20z = 1500$ and $12x + 20y + 14z = 1600$ replaced by the inequalities $10x + 15y + 20z \leqslant 1500$ and $12x + 20y + 14z \leqslant 1600$, just to see if there is much improvement in the optimal solution. The new program is below. Notice lines 44 and 45 reverse the less than or equal constraints so that it will take a false result in both lines to get to the profit function in line 50. And this, of course, will only happen if both less than or equal constraints in the problem hold.

```
5     REM FIXED COST MAXIMIZATION
6     REM WITH INEQUALITY CONSTRAINTS
10    B=-999999
20    FOR X=0 TO 133
30    FOR Y=0 TO 80
40    FOR Z=0 TO 75
44    IF 10*X+15*Y+20*Z > 1500 THEN 210
45    IF 12*X+20*Y+14*Z > 1600 THEN 210
50    P=71*X+52*Y+29*Z-390
60    IF X = 0 THEN 100
70    IF Y = 0 THEN 120
80    IF Z = 0 THEN 140
90    GO TO 150
100   P=P+200
110   GO TO 70
120   P=P+110
130   GO TO 80
140   P=P+80
150   IF P > B THEN 170
160   GO TO 210
170   A1=X
```

```
180  A2=Y
190  A3=Z
200  B=P
210  NEXT Z
220  NEXT Y
230  NEXT X
240  PRINT A1,A2,A3
250  PRINT B
260  STOP
270  END

133          0          0
9963
```

Now the optimum has increased to \$9963 by making only Product A (133 units is maximal capacity). So management might consider a change or discuss with the union the redeployment of some workers. Again, the problem could be modified to see if hiring one or two new workers and/or changing some workers' assignments to different departments would increase profit (as long as these moves did not increase production past demand for the product).

EXAMPLE 3.6

Consider the problem to maximize $P = 3x + 6y + 5z + xy$ subject to $x \geqslant 0, y \geqslant 0, z \geqslant 0, x + 2y + 10z = 80$, and $3x + y + 6z = 100$.

By solving the system of equations first, we have

$$\begin{array}{rrrr} 1 & 2 & 10 & 80 \\ 3 & 1 & 6 & 100 \\ 1 & 2 & 10 & 80 \\ 0 & -5 & -24 & -140 \\ 1 & 2 & 10 & 80 \\ 0 & 1 & 24/5 & 28 \\ 1 & 0 & 2/5 & 24 \\ 0 & 1 & 24/5 & 28 \end{array}$$

Therefore, arbitrary z produces:

$$x = 24 - .4z$$

$$y = 28 - 4.8z$$

Actually, $0 \leqslant z \leqslant 5.833$ to keep all variables positive (from $28 = 4.8z$, therefore $z = 28/4.8 = 5.833$). We now write a BASIC program to solve our previously stated maximization problem. Note that lines 20, 30, 40, and 50 deal with the solution of the system of equations. The structure of the program is the same as in previous examples.

```
5     REM NONLINEAR OPTIMIZATION
6     REM WITH EQUATIONS
10    B=-999999
20    FOR I=0 TO 5833
30    Z=I/1000.
40    X=24-.4*Z
50    Y=28-4.8*Z
60    P=3*X+6*Y+5*Z+X*Y
70    IF P > B THEN 90
80    GO TO 130
90    B=P
100   X1=X
110   Y1=Y
120   Z1=Z
130   NEXT I
140   PRINT X1,Y1,Z1
150   PRINT B
160   STOP
170   END

24              28        0
912
```

Therefore, when $x = 24$, $y = 28$, and $z = 0$, P takes its maximum value of 912.

EXAMPLE 3.7

Now, let us take the previous problem and try to minimize the P equation subject to the same constraints:

```
4     REM PREVIOUS PROBLEM MINIMIZED
5     REM NONLINEAR OPTIMIZATION
6     REM WITH EQUATIONS
10    B=999999
20    FOR I=0 to 5833
30    Z=I/1000.
40    X=24-.4*Z
50    Y=28-4.8*Z
```

```
60   P=3*X+6*Y+5*Z+X+Y
70   IF P < B THEN 90
80   GO TO 130
90   B=P
100  X1=X
110  Y1=Y
120  Z1=Z
130  NEXT I
140  PRINT X1,Y1,Z1
150  PRINT B
160  STOP
170  END

21.6668          1.60000E-03      5.83300
94.2097
```

Therefore, $x = 21.6668$, $y = .0016$, and $z = 5.833$ yields a minimum P of 94.2097.

EXAMPLE 3.8

Now let us try to maximize $P = 2x + 9y + 14z + 2zx - z^2$ subject to $x \geqslant 0, y \geqslant 0, z \geqslant 0$, and $x + 2y^2 + 10z = 80$ and $3x + y + 6z = 100$.

We rewrite the equations as

$$x + 0y + 2y^2 + 10z = 80$$

$$3x + y + 0y^2 + 6z = 100$$

and solve this system (keeping in mind that the second variable squared must equal the third variable). We get the following:

$$\begin{array}{rrrrr} 1 & 0 & 2 & 10 & 80 \\ 3 & 1 & 0 & 6 & 100 \\ 1 & 0 & 2 & 10 & 80 \\ 0 & 1 & -6 & -24 & -140 \end{array}$$

Therefore

$$x + 0y + 2y^2 + 10z = 80$$

$$0x + y - 6y^2 - 24z = -140$$

For arbitrary y^2 and z we have

$$x = 80 - 10z - 2y^2$$

$$y = -140 + 24z + 6y^2$$

But we must be careful to insure that

$$y = -140 + 24z + 6y^2 \text{ with } y = y^2$$

Therefore, we should state for arbitrary y^2, take the square root of y^2, and assign this value to y in the second equation which we have solved for z as follows:

$$z = \frac{y - 6y^2 + 140}{24}$$

Then $x = 80 - 10z - 2y^2$

Now from $x + 2y^2 + 10z = 80$ we see that $y^2 \leqslant 40$ if $x \geqslant 0$ and $z \geqslant 0$. So we will write a program varying y^2 from 0 to 40 (by increments of .01, for example) and in each case produce x, y, and z from it. As long as x, y, y^2, and z are all nonnegative, we will have the program evaluate the function and check the feasible solution for optimality in the usual fashion. The program follows:

```
5     REM NONLINEAR OPTIMIZATION WITH
6     REM EQUATIONS THAT ARE NONLINEAR
10    B=-999999
20    FOR I=0 TO 4000
30    Y2=I/100.
40    Z=(SQR(Y2)-6*Y2+140)/24.
50    X=80-10*Z-2*Y2
60    IF Z < -.000001 THEN 160
70    IF X < -.000001 THEN 160
80    P=2*X+9*SQR(Y2)+14*Z+2*Z*X-Z**2
90    IF P > B THEN 110
100   GO TO 160
110   A1=X
120   A2=SQR(Y2)
130   A3=Y2
140   A4=Z
150   B=P
160   NEXT I
170   PRINT A1,A2,A3,A4
180   PRINT B
190   STOP
200   END

21.5910          .565685          .320000      5.77690
345.235
```

Therefore, $x = 21.5910$, $y = .565685$, $y^2 = .320000$, and $z = 5.77690$ yields a maximum P of 345.235.

EXAMPLE 3.9

Now let us alter the above program in order to find the minimum of the system. Note that this time we will search by increments of .001 (line 30) instead of .01 (just to demonstrate that more accuracy is possible if necessary). The program is below:

```
5     REM NONLINEAR OPTIMIZATION WITH
6     REM EQUATIONS THAT ARE NONLINEAR
7     REM MINIMIZE THE FUNCTION
10    B=999999
20    FOR I=0 TO 40000
30    Y2=I/1000.
40    Z=(SQR(Y2)-6*Y2+140)/24.
50    X=80-10*Z-2*Y2
60    IF Z < -.000001 THEN 160
70    IF X < -.000001 THEN 160
80    P=2*X+9*SQR(Y2)+14*Z*X-Z**2
90    IF P < B THEN 110
100   GO TO 160
110   A1=X
120   A2=SQR(Y2)
130   A3=Y2
140   A4=Z
150   B=P
160   NEXT I
170   PRINT A1,A2,A3,A4
180   PRINT B
190   STOP
200   END

31.6950        4.91447        24.1520        1.02851E-04
107.628
```

Therefore, $x = 31.6950$, $y = 4.91447$, $y^2 = 24.1520$, and $z = .000102851$ yields a minimum P of 107.628.

These types of problems (nonlinear objective functions with nonlinear equation constraints) can occur in industrial blending problems, oil refinery problems, or chemical yield systems. They are difficult to solve theoretically, but present little trouble when done with our computer approach.

EXAMPLE 3.10

A farm silo of cylindrical shape with a half-dome top is to be built to have a capacity of 300 cubic meters. The height of the cylinder is to be between 10 and 80 meters. The radius of the cylinder (and half-dome top) is to be greater than or equal to 1.

Cylinder volume and dome volume equal 300 cubic meters. Therefore,

$$\pi r^2 h + .5\tfrac{4}{3}\pi r^3 = 300$$

Solving for r, we see that r must be less than or equal to 3.1 meters.

We will vary r by hundredths of a meter. The sides, floor, and roof cost \$20, \$12, and \$30 per square meter, respectively. Therefore, we seek to minimize $C = 2\pi\ rh(20) + \pi\ r^2(12) + .5(4\pi r^2) \times (30)$ subject to $1 \leqslant r \leqslant 3.1$, $10 \leqslant h \leqslant 80$, and $\pi r^2 h + .5\tfrac{4}{3}\pi r^3 = 300$.

The program follows:

```
6     REM OF COST OF BUILDING FARM
7     REM SILO SUBJECT TO CONSTRAINTS
10    B=999999
20    FOR R=1 TO 3.1 STEP .01
28    H1=300-.666667*3.14159*R**3
29    H2=3.14159*R**2
30    H=H1/H2
40    IF H < 10 THEN 120
50    IF H > 80 THEN 120
58    C1=40*3.14159*R*H
59    C2=12*3.14159*R**2+60*3.14159*R**2
60    C=C1+C2
70    IF C < B THEN 90
80    GO TO 120
90    A1=R
100   A2=H
110   B=C
120   NEXT R
130   PRINT A1,A2
140   PRINT B
150   STOP
160   END

2.83000         10.0367
5380.90
```

The printout is:

2.830 10.037 5380.898

Therefore, $r = 2.83$, $h = 10.037$, and $c = 5380.90$. So the silo should have a radius of 2.83 meters and a height of 10.037 meters to yield a minimum cost of $5380.90.

EXAMPLE 3.11

A container with rectangular sides and a base and top that are equilateral triangles is to be made to hold 2000 cubic centimeters. The surface area is to be minimized. This will minimize the cost since all five sides are made from the same material. The side of the triangle (s) must be between 2 and 20 centimeters and the length (l) of the rectangular side must be between 10 and 100 centimeters. The height of the triangle (h) equals $S\sqrt{3}/2$.

Therefore, we seek to minimize $A = .5sh \times (2) + 3sl$ subject to $2 \leqslant s \leqslant 20$, $10 \leqslant l \leqslant 100$, and $.5shl = 2000$.

We vary s by hundredths of a centimeter in the program:

```
5     REM THREE VARIABLE MINIMIZATION
6     REM OF COST OF MAKING TRIANGULAR
7     REM CONTAINER SUBJECT TO CONSTRAINTS
10    B=999999
20    FOR S=2 TO 20 STEP .01
30    H=SQR(3.)/2*S
40    L=2000/(.5*S*H)
50    C=S*H+3*S*L
60    IF C < B THEN 80
70    GO TO 120
80    A1=S
90    A2=H
100   A3=L
110   B=C
120   NEXT S
130   PRINT A1,A2,A3
140   PRINT B
150   STOP
160   END

20.0000          17.3205          11.5470
1039.23
```

The printout is:

20.000 17.320 11.547 1039.23

Therefore, $s = 20$, $h = 17.32$, $l = 11.547$, and $A = 1039.23$. With equilateral triangle sides of 20 centimeters each and rectangular sides of length 11.547 centimeters each, we minimize the surface area of the container under our constraints.

EXAMPLE 3.12

Minimize the area of a rectangular container of volume 3000 cubic centimeters, if the height (h) must be between 2 and 20 centimeters and the width (w) must be between 3 and 20 centimeters.

Therefore, minimize $A = 2hw + 4hl$ subject to $2 \leqslant h \leqslant 20$, $3 \leqslant w \leqslant 20$, and $hwl = 3000$.

From the program we can see that a height of 2 centimeters, a width of 20 centimeters, and a length of 75 centimeters will minimize the surface area.

```
5     REM THREE VARIABLE MINIMIZATION
6     REM OF AREA OF RECTANGULAR CONTAINER
10    B=999999
20    FOR H=2 TO 20 STEP .1
30    FOR W=3 TO 20 STEP .1
40    L=3000./(H*W)
50    A=2*H*W+4H*L
60    IF A < B THEN 80
70    GO TO 120
80    A1=H
90    A2=W
100   A3=L
110   B=A
120   NEXT W
130   NEXT H
140   PRINT A1,A2,A3
150   PRINT B
160   STOP
170   END

2                  20.0000              75.0000
680.000
```

The printout is:

2.000 20.000 75.000 680.000

The demand curve (price-quantity curve) is $P(z) = 6 + 1200/q$. The cost curve is $7 + .8q$. And no more than 100 units per day can be produced.

Therefore, we seek to maximize $P = 6q + 1200 - 7q - .8q^2$ subject to $0 \leqslant q \leqslant 100$.

From the printout we can see that the company should consider stopping production of this product unless it can improve its price and/or cost curves.

```
5     REM ONE VARIABLE PROFIT MAXIMIZATION
6     REM TAKING DEMAND CURVE AND COSTS INTO ACCOUNT
10    B=-999999
20    FOR Q=0 TO 100
30    P=6*Q+1200-7*Q-.8*Q**2
40    IF P > B THEN 60
50    GO TO 80
60    A1=Q
70    B=P
80    NEXT Q
90    PRINT A1,B
100   STOP
110   END

0              1200
```

The cost curve is $\$22 + \$.85q + \$.02q^{1.4}$. The price curve is $\$200 - \$.6q$. Also, we can produce no more than 1000 units per day.

Therefore, we seek to maximize Profit $= (200 - .6q)q - (22 + .85q + .02q^{1.4})q$ subject to $0 \leqslant q \leqslant 1000$ units per day.

The printout below tells us that 57 units per day will maximize the profit.

```
5     REM ONE VARIABLE PROFIT MAXIMIZATION
6     REM TAKING DEMAND CURVE AND COSTS INTO ACCOUNT
10    B=-999999
20    FOR Q=0 TO 1000
30    P=(200-.6*Q)*Q-(22.+.85*Q+.02*Q**1.4)*Q
40    IF P > B THEN 60
50    GO TO 80
60    A1=Q
70    B=P
80    NEXT Q
90    PRINT A1,B
100   STOP
110   END

57             5107.51
```

EXAMPLE 3.13

We seek to maximize $N = 25L^{.4}C^2$, where N is the number of units of production and L and C are labor and capital. Each laborer costs 500 cents per time period. There must be at least 10 laborers. Capital is spent in units of 100 cents. There is a \$10,000 budget available. So $500L + 100C = 1{,}000{,}000$ cents.

The printout below shows that \$8333 and 333 laborers will maximize production:

```
5     REM TWO VARIABLE MAXIMIZE
6     REM PRODUCTION AS A FUNCTION
7     REM OF LABOR AND CAPITAL
8     REM UNDER CONSTRAINTS
10    B=-999999
20    FOR C=0 to 10000
30    L=(1000000-100*C)/500.
40    N=25*L**.4*C**2
50    IF N > B THEN 70
60    GO TO 100
70    A1=C
80    A2=1
90    B=N
100   NEXT C
110   PRINT A1,A2
120   PRINT B
130   STOP
140   END

8333                333.400
1.77308E+10
```

EXAMPLE 3.14

Minimize the cost function $C = 15 + 7x^2 + 3x + 9xy^2$ subject to $0 \leqslant x \leqslant 100$, $y \geqslant 0$, and $xy = 2400$. The program is below:

```
5     REM TWO VARIABLE MINIMIZE COST
6     REM FUNCTION UNDER CONSTRAINTS
7     B=999999
10    FOR X=1 TO 100
20    Y=2400./X
30    C=15+7*X**2+3*X+9*X*Y**2
40    IF C < B THEN 60
```

```
50   GO TO 90
60   A1=X
70   A2=Y
80   B=C
90   NEXT X
100  PRINT A1,A2
110  PRINT B
120  STOP
130  END

100              24
588715
```

Therefore, $x = 100$ and $y = 24$ yields the minimum C of 588,715.

EXAMPLE 3.15
The advertising department in the K Company is trying to develop the most effective advertising strategy given their budgetary and other constraints. They have five possible ads to choose from. They are:

Ad 1: Sponsoring television show A (limit of at most 10 times)

Ad 2: Sponsoring television show B (limit of at most 15 times)

Ad 3: Sponsoring radio show C (limit of at most 8 times)

Ad 4: A one-page ad in magazine T (limit of at most 4 times)

Ad 5: A Sunday newspaper ad of four pages (limit of at most twice)

These limits are all in the time period of the next 8 months. The budget for the next 8 months is $1,000,000.

The costs of the ads and their profit ratings relative to each other (a rating of 2 is twice as productive saleswise as a 1, etc.) are given below:

Ad	Cost per unit ($)	Profit rating	Number of units
1	150,000	14	x_1
2	90,000	8	x_2
3	10,000	3	x_3
4	5,000	2.4	x_4
5	15,000	3.9	x_5

Management wants to do at least one Sunday newspaper ad (Ad 5) and at least 2 sponsorships of television show 2 (Ad 2). What advertising strategy will maximize the profit rating subject to the constraints?

We seek to maximize $P = 14x_1 + 8x_2 + 3x_3 + 2.4x_4 + 3.9x_5$ subject to $0 \leqslant x_1 \leqslant 10$, $2 \leqslant x_2 \leqslant 15$, $0 \leqslant x_3 \leqslant 8$, $0 \leqslant x_4 \leqslant 4$, $1 \leqslant x_5 \leqslant 2$, and $150{,}000x_1 + 90{,}000x_2 + 10{,}000x_3 + 5000x_4 + 15{,}000x_5 \leqslant 1{,}000{,}000$. The program follows:

```
4    REM FIVE VARIABLE
5    REM ADVERTISING STRATEGY
6    REM PROBLEM
10   B=-999999
20   FOR X1=0 TO 10
30   FOR X2=2 TO 15
40   FOR X3=0 TO 8
50   FOR X4=0 TO 4
60   FOR X5=1 TO 2
70   IF 150*X1+90*X2+10*X3+5*X4+15*X5 > 1000 THEN 170
80   P=14*X1+8*X2+3*X3+2.4*X4+3.9*X5
90   IF P > B THEN 110
100  GO TO 170
110  A1=X1
120  A2=X2
130  A3=X3
140  A4=X4
150  A5=X5
160  B=P
170  NEXT X5
180  NEXT X4
190  NEXT X3
200  NEXT X2
210  NEXT X1
220  PRINT A1,A2,A3,A4,A5
230  PRINT B
240  STOP
250  END
```

```
4           3           8           4           2
121.400
```

Therefore, the K Company should sponsor television show A 4 times, show B 3 times, radio show C 8 times, 4 ads in magazine T and the Sunday newspaper ad twice to yield a profit rating of 121.4.

Exercise: Modify the above program to print out all the answers that have a profit rating greater than 116. This would give the company several options to choose from and yet have about the same effectiveness exposure.

EXAMPLE 3.16

The Medford Insurance Company wants to invest no more than $900,000 in blocks of $100,000 in five different securities. Their returns are listed below:

Security	Return(%)	Amount available ($)
1	8	200,000
2	12	300,000
3	15	200,000
4	7	100,000
5	9	100,000

Government regulations state that at least 50% of the investment must be in securities 1, 4, or 5 because their risks are lower and hence will not jeopardize the policyholders' reserves. In addition, company policy says that at least $100,000 of each security must be purchased. What investment plan will maximize the return under these conditions? The program follows:

```
4     REM FIVE VARIABLE FINANCIAL
5     REM PLANNING PROBLEM
6     REM INSURANCE REGULATED
7     REM INVESTMENT
10    B=-999999
20    FOR X1=0 TO 200000 STEP 100000
30    FOR X2=0 TO 300000 STEP 100000
40    FOR X3=0 TO 200000 STEP 100000
50    FOR X4=0 TO 100000 STEP 100000
60    FOR X5=0 TO 100000 STEP 100000
70    IF X1+X4+X5 < .5*(X1+X2+X3+X4+X5) THEN 170
80    P=.08*X1+.12*X2+.15*X3+.07*X4+.09*X5
90    IF P > B THEN 110
100   GO TO 170
110   A1=X1
120   A2=X2
130   A3=X3
140   A4=X4
150   A5=X5
160   B=P
170   NEXT X5
180   NEXT X4
190   NEXT X3
200   NEXT X2
```

```
210  NEXT X1
220  PRINT A1,A2,A3,A4,A5
230  PRINT B
240  STOP
250  END

200000     200000         200000        100000    100000
85999
```

Therefore, the company should buy \$200,000 worth of Security 1, \$200,000 of Security 2, \$200,000 of Security 3, \$100,000 of Security 4, and \$100,000 of Security 5. The return will be \$85,999.

EXAMPLE 3.17

The Pacific Palisades Company has two factories and two warehouses. The supplies of the factories, the demands of the warehouses, and the unit shipping costs are shown in the table below:

		Warehouse 1	Warehouse 2	Supply
Factory	1	\$50	\$65	400
	2	\$70	\$40	500
	Demand	300	600	

There are the following fixed charges which are incurred only if a non-zero amount is shipped from factory to warehouse:

Factory 1	to	Warehouse 1	\$ 800
Factory 2	to	Warehouse 2	\$2000
Factory 2	to	Warehouse 1	\$1500
Factory 2	to	Warehouse 2	\$ 600

We seek the shipment pattern that will minimize the cost. This can be summarized as minimize $C = 50x_{11} + 800y_{11} + 65x_{12} + 2000y_{12} + 70x_{21} + 1500y_{21} + 40x_{22} + 600y_{22}$ subject to $x_{11} \geqslant 0$, $x_{12} \geqslant 0$, $x_{21} \geqslant 0$, $x_{22} \geqslant 0$, $x_{11} + x_{12} = 400$, $x_{21} + x_{22} = 500$, $x_{11} + x_{21} = 300$, $x_{12} + x_{22} = 600$, and

$y_{11} = 1$ if $x_{11} > 0$ and $y_{11} = 0$ if $x_{11} = 0$

$y_{12} = 1$ if $x_{12} > 0$ and $y_{12} = 0$ if $x_{12} = 0$

$y_{21} = 1$ if $x_{21} > 0$ and $y_{21} = 0$ if $x_{21} = 0$
$y_{22} = 1$ if $x_{22} > 0$ and $y_{22} = 0$ if $x_{22} = 0$

First we will solve the system of equations:

$$
\begin{aligned}
x_{11} + x_{12} \qquad\qquad\qquad &= 400 \\
x_{12} + \qquad\quad x_{22} &= 600 \\
x_{21} + x_{22} &= 500 \\
x_{11} \quad + \quad x_{21} \qquad\quad &= 300
\end{aligned}
$$

Using the row diagonalization method discussed earlier in this chapter we get:

1	1	0	0	400
0	1	0	1	600
0	0	1	1	500
1	0	1	0	300
1	1	0	0	400
0	1	0	1	600
0	0	1	1	500
0	-1	1	0	-100
1	0	0	-1	-200
0	1	0	1	600
0	0	1	1	500
0	0	1	1	500
1	0	0	-1	-200
0	1	0	1	600
0	0	1	1	500
0	0	0	0	0

Therefore, the solution is:

$$
\begin{aligned}
x_{11} - x_{22} &= -200 \\
x_{12} + x_{22} &= 600 \\
x_{21} + x_{22} &= 500
\end{aligned}
$$

or for $200 \leqslant x_{22} \leqslant 500$, then

$$x_{11} = x_{22} - 200$$
$$x_{12} = 600 - x_{22}$$
$$x_{21} = 500 - x_{22}$$

Keep in mind the fixed-charge terms while looking at the program below:

```
4    REM FOUR VARIABLE FIXED CHARGE
5    REM TRANSPORTATION PROBLEM
6    REM X1 IS AMOUNT SHIPPED FROM
7    REM FACTORY 1 TO WAREHOUSE 1
8    REM X2 IS AMOUNT SHIPPED FROM
9    REM FACTORY 1 TO WAREHOUSE 2
10   REM Y1 IS AMOUNT SHIPPED FROM
11   REM FACTORY 2 TO WAREHOUSE 1
12   REM Y2 IS AMOUNT SHIPPED FROM
13   REM FACTORY 2 TO WAREHOUSE 2
14   B=999999
20   FOR Y2=200 TO 500
30   X1=Y2-200
40   X2=600-Y2
50   Y1=500-Y2
60   C=50*X1+65*X2+70*Y1+40*Y2
70   C=C+800+2000+1500+600
80   IF X1=0 THEN 130
90   IF X2=0 THEN 150
100  IF Y1=0 THEN 170
110  IF Y2=0 THEN 190
120  GO TO 200
130  C=C-800
140  GO TO 90
150  C=C-2000
160  GO TO 100
170  C=C-1500
180  GO TO 110
190  C=C-600
200  IF C < B THEN 220
210  GO TO 270
220  A1=X1
230  A2=X2
240  B1=Y1
250  B2=Y2
260  B=C
270  NEXT Y2
280  PRINT A1,A2
```

```
290  PRINT B1,B2
300  PRINT B
310  STOP
320  END

300          100
0            500
44900
```

The printout is:

300 100 0 500 44900

Therefore, the company should ship 300 units from Factory 1 to Warehouse 1, 100 units from Factory 1 to Warehouse 2, and 500 units from Factory 2 to Warehouse 2 in order to minimize the cost.

EXERCISES

3.1 Maximize $P = 6x_1 + 9x_2 + 7x_3 + 10x_4$ subject to all $x_i \geqslant 0$, $x_1 + 2x_2 + 3x_3 + x_4 \leqslant 500{,}000$, and $5x_1 + 4x_2 + 2x_4 \leqslant 600{,}000$.

Hint. Use a Monte Carlo simulation and then focus in and search for the true optimal solution.

3.2 Minimize $C = 2x_1 + 3x_2 + 2.6x_3$ subject to $x_i \geqslant 0$ for all i, $x_1 + 2x_2 + x_3 = 1000$, $2x_1 + 3x_2 + 4x_3 = 3000$, and $x_1 + x_2 + x_3 \leqslant 1000$.

Hint. Solve the system of equations first and remember to bound the solutions so that all $x_i \geqslant 0$.

3.3 Maximize $P = 2x_1 + 3x_2 + 2.6x_3$ subject to $x_i \geqslant 0$ for all i, $x_1 + 2x_2 + x_3 = 1000$, $2x_1 + 3x_2 + 4x_3 = 3000$, and $x_1 + x_2 + x_3 \leqslant 1000$.

3.4 Maximize $P = 2x_1 + 3x_2 + 2.6x_3 + 5x_2x_3$ subject to $x_i \geqslant 0$ for all i, $x_1 + 2x_2 + x_3 = 1000$, $2x_1 + 3x_2 + 4x_3 = 3000$, and $x_1 + x_2 + x_3 \leqslant 1000$.

3.5 The Cedarville Metal Machining Co. buys raw metal plates and machines them into Grade A finished metal, Grade B, and Grade C. The amount of time each unit of A, B, and C spends in the six machining departments is listed below. Grade A returns \$100 profit per unit, B a \$61 profit, and C a \$54 profit. Find the maximum profit solution. (Hint: Do a simulation.)

Time required per unit				
A	B	C	Departments	Hours available
1	2	1	Turning	6,000
3	1	2	Planing	5,000
4	1	2	Drilling	10,000
1	1	1	Grinding	4,000
.6	1	.2	Milling	3,000
.4	2	.7	Lapping	8,000

3.6 Substitute the solution into the constraints and go to work reassigning workers to increase production and profit.

3.7 Helberg Manufacturing Company is trying to plan an advertising campaign. They want to allocate $300,000 among the following four types of ads in order to maximize their profit rate:

Ad	Type	Profit rating	Cost per unit ($)
1	Selected TV ads	10.6	50,000
2	Newspaper campaign	11.9	55,000
3	Full-page ad in Magazine A	5.2	22,000
4	Half-page ad in Magazine B	2.9	10,000

Management has decreed that at least 3 different types of ads must be used and at least 50% of the budget must be spent on ads 2 and 3. Therefore

1. Find the solution that maximizes the profit rating.
2. Find the five next best solutions to Helberg's ad campaign so that management will have a choice.
3. Optimize the Helberg ad campaign after dropping the constraints.
4. Find the next best solution to problem 3.

3.8 The Bergen Bay School District has just been created by the Town of Hawkewood. People have been moving into Bergen Bay on Hawkewood's west side at such a rapid rate that the new school district was formed. The city council is trying to select the site for the new Bergen Bay School. The district has 8 regions (listed along with number of school children). As with most school districts, a political fight looms over where to build the new school. The city council is opting for the region that will minimize the total distance traveled to school by all children. This will take the council off the hook and provide a good solution. Find the minimum distance solution using the chart on page 69. Note that all one needs to do is sum the miles times number of children for the eight different locations and corresponding mileages and take the minimum one.

Distance in miles from central part of each region to proposed school location in each other region

Number of students	Ter. Bay.	Thund. Bay	E. Riv.	Rock H.	Echo B.	Pop. Bl.	Rhine.	Wood.
85 - Terrace Bay		.7	.4	1.2	.9	1.3	4	2.1
114 - Thunder Bay			.9	1.1	.6	2.0	1.5	1.7
79 - Eagle River				.8	.9	1.4	1.6	1.9
58 - Rock Harbor					.9	.8	.7	2.6
102 - Echo Bay						1.7	1.1	1.0
83 - Poplar Bluff							1.6	2.0
49 - Rhinelander								1.4
66 - Woodruff								

Note that in a facilities location problem like this that the computer could accommodate up to about ten thousand locations and sum the entries and minimize easily (as long as someone keyed in the data).

The author believes that these facilities location problems (industrial ones, too) have real potential for cutting costs, finding a good solution, helping the most people, and, last but not least, as a political tool for state, local, and/or federal government to take the pressure off of them in political battles that develop over facilities locations.

3.9 The Grant's Pass Company is making three products in departments 1 and 2. Products X and Z both return \$8 profit per unit, while Product Y earns \$10 profit per unit. One unit of X takes 1 hour in Department 1 and .5 hours in Department 2. One unit of Y takes 1 hour in each department. One unit of Z takes .5 hours in each Department. There are 1000 worker-hours available in Department 1 and 800 worker-hours available in Department 2.

This problem is complicated by the fact that these products are being made by summer students earning money for school. Management has agreed that no one is to be standing around idle. Therefore, on this first set-up for production (until management can adjust the right number of workers in each department) all 1000 and 800 worker-hours are to be used. Hence, we want to maximize $P = 8x + 10y + 8z$ subject to $x \geqslant 0, y \geqslant 0, z \geqslant 0, x + y + .5z = 1000$, and $.5x + y + .5z = 800$.

Hint: Solve the system of equations and then arrange to deal only with the answers where x, y, and z are all nonnegative.

3.10 The Churchill Falls Company has a blending program. Coffee Blend and Best Roast Coffee are to be made from Columbia Best Coffee (\$2 per pound) and Brazilian Coffee (\$1.85 per pound). 10,000 pounds of Coffee Blend and 15,000 pounds of Best Roast are to be made. At least 30% of Best Roast must be Columbia Best Coffee, and not more than 65% of Coffee Blend can be Brazilian Coffee. Coffee Blend sells for \$3 per pound and Best Roast for \$3.30. We seek to maximize profit. Let x_{11} be the number of units of Columbia Best put into Coffee Blend. Let x_{12} be the number of units of Columbia Best put into Best Roast. Let x_{21} be the number of units of Brazilian Coffee put into Coffee Blend, and let x_{22} be the number of units of Brazilian put into Best Roast.

Hint: Solve the system of equations first.

3.11 The Harbor Springs Firefighters are about to join forces with the town's rescue workers (paramedics, ambulance service, etc.). Both the firemen and rescue people are to be housed at the Lake Michigan Station. They start work at either 6:00 A.M., 12:00 noon, 6:00 P.M., or 12:00 midnight and work 12-hour shifts. From studying past statistics they believe that the scheduling should meet the following requirements.

Time period	Minimum no. of firemen needed	Minimum no. of rescue people
6:00 A.M. to noon	6	3
12:00 noon to 6:00 P.M.	10	3
6:00 P.M. to midnight	11	4
Midnight to 6:00 A.M.	7	5

Firemen and rescue workers are paid at the same rate. They both earn time and a half for hours worked between 6:00 P.M. and 6:00 A.M. In addition, many firemen are also qualified rescue people and vice versa. These people earn a 15% bonus on top of regular pay rate (it is 15% of the base pay whether there are overtime hours or not).

The city council is under a budget strain and they are also a little short of qualified people. Therefore, they are going to try a scheduling plan where they count the people who are double qualified as one fireman and one rescue worker when they are on duty. Given that there are five such people available each day, find the minimum cost solution to Harbor Springs' scheduling program. *Hint:* Let x_i be the number of firemen starting work in shift i, and y_i be the similar thing for rescue workers. Let z_i be for the double qualified. Solve the resultant system of 8 equations and 12 variables (plus the $z_1 + z_2 + z_3 + z_4 = 5$) and go to work on the cost equation with a BASIC program.

3.12 The Republic Fire Department has a scheduling problem. Their firemen start work at 6:00 A.M., 12:00 noon, 6:00 P.M., or 12:00 midnight and work for 12 hours straight. By studying the past records of fires in Republic's territory, they believe that they should have firemen available according to the following schedule:

Time period	No. of firemen needed
6:00 A.M. to noon	At least 10
12:00 noon to 6:00 P.M.	At least 15
6:00 P.M. to midnight	At least 20
Midnight to 6:00 A.M.	At least 18

Formulate this integer programming problem to minimize the number of firemen used and still meet the scheduling conditions. A four-loop BASIC program should solve it easily.

Suggested Reading

Brown, K.S. and ReVelle, J.B. *Quantitative Methods for Managerial Decisions*. Reading, Mass.: Addison-Wesley, 1978.

Fuori, William. *Introduction to the Computer, the Tool of Business*. Englewood Cliffs, N.J.: Prentice-Hall, 1973.

Heinze, David. *Management Science, Introductory Concepts and Applications*. Cincinnati: Southwestern Co., 1978.

Smith, Derrick. *Linear Programming Models in Business*. Stockport, England: Polytech Publishers, 1973.

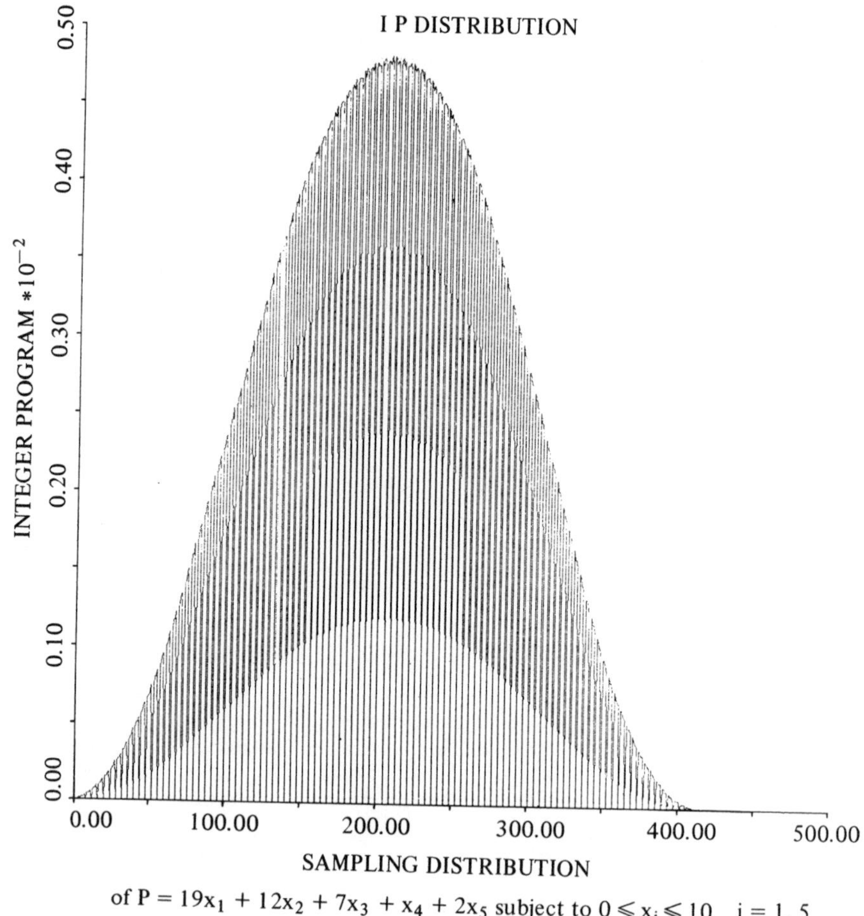

of $P = 19x_1 + 12x_2 + 7x_3 + x_4 + 2x_5$ subject to $0 \leqslant x_i \leqslant 10 \quad i = 1, 5$

FIGURE 4.1 is another sampling distribution of an integer programming problem. Because the distributions are connected, the odds are overwhelming that we will get a nearly optimal solution by looking at a random sample of thousands of feasible solutions and taking the maximal one. This allows us to find the solution to problems that were previously considered unsolvable.

CHAPTER 4

Inventory and Econometric Models

Inventory problems deal with fixed costs, holding costs, ordering costs, and various constraints. These factors usually interact to produce a nonlinear model that one wishes to minimize.

Econometrics tries to discover and analyze the many and varied factors that contribute to a company's or a country's economic health. These factors usually interact, giving rise to nonlinear models that one wants to optimize.

This chapter presents some examples of both types of models.

Inventory Models

EXAMPLE 4.1

The Kalispell Company is trying to keep their inventory costs under control for Product T. They use 12,000 units of T a year. It costs $1000 per unit per year to store Product T, $82 per order to order q units of T. Therefore, the inventory cost equation for Product T can be written as $C = 1000Q/2 + (82)12{,}000/Q$, where $Q/2$ is the average number of units of Product T in the warehouse at any one given time (times the $1000 gives the annual holding or storage cost), and $12{,}000/Q$ is the number of orders per year (times $82 per order is the annual ordering cost).

This is a standard economic order quantity or EOQ model. Below is a BASIC program to minimize this inventory cost equation:

```
5     REM STANDARD INVENTORY COST
6     REM MINIMIZATION WITH Q
7     REM THE NUMBER OF UNITS
8     REM IN EACH ORDER
10    B=999999
20    FOR Q=1 TO 12000
30    C=1000*Q/2+82*12000/Q
40    IF C < B THEN 60
50    GO TO 80
60    Q1=Q
70    B=C
80    NEXT Q
90    PRINT Q1
100   PRINT B
110   STOP
120   END

44
44363.6
```

Therefore, the company should order 44 units of T each time it places an order to yield a minimum cost of $44,363.60 per year.

Further, the Kalispell Company has the inventory cost equation $C = 1500Q/2 + (100)(1400/Q)$ for Product JJ. The BASIC program to minimize this cost equation follows:

```
5     REM STANDARD INVENTORY COST
6     REM MINIMIZATION WITH Q
7     REM THE NUMBER OF UNITS
8     REM IN EACH ORDER
10    B=999999
20    FOR Q=1 TO 1400
30    C=1500*Q/2+100*1400/Q
40    IF C < B THEN 60
50    GO TO 80
60    Q1=Q
70    B=C
80    NEXT Q
90    PRINT Q1
```

```
100  PRINT B
110  STOP
120  END

14
20500
```

Therefore, they should order Product JJ in lots of 14.

EXAMPLE 4.2

The BASIC EOQ model above assures a so-called infinite replenishment rate, namely, that the entire order arrives at once. However, Cook and Russell discuss an EOQ with a finite replenishment rate where, for instance, the company is producing for inventory.[1] In this case, the order arrives at a specified rate and not immediately with the first shipment. This has the effect of lowering the average holding cost somewhat. An example would be minimize the cost equation $C = .18Q(1 - 15{,}000/100{,}000)/2 + (20)15{,}000/Q$, where Q is the quantity ordered each time, 15,000 is the total demand for the year, 100,000 is the replenishment rate per year, \$.18 is the holding cost per unit per year, and \$20 is the ordering cost per order. Even though this is a more complicated equation mathematically, it is just as easy to solve in BASIC. The program follows:

```
5    REM STANDARD INVENTORY COST
6    REM MINIMIZATION WITH Q
7    REM THE NUMBER OF UNITS
8    REM IN EACH ORDER, BUT
9    REM WITH FINITE REPLENISHMENT RATE
10   B=999999
20   FOR Q=1 TO 100000
30   C=.18*Q*(1-15000/100000)/2+20*15000/Q
40   IF C < B THEN 60
50   GO TO 80
60   Q1=Q
70   B=C
80   NEXT Q
90   PRINT Q1
```

[1]Thomas Cook and Robert Russell, *Introduction to Management Science* (Englewood Cliffs, N.J.: Prentice-Hall, 1977), p. 390

```
100  PRINT B
110  STOP
120  END

1980
302.985
```

So each reorder should request 1980 units to minimize the inventory cost.

EXAMPLE 4.3

Another finite replenishment rate problem would be to minimize $C = .125Q(1 - 10{,}000/125{,}000)/2 + (25)10{,}000/Q$. The program is:

```
5     REM ONE VARIABLE INVENTORY
6     REM COST MINIMIZATION PROBLEM
7     REM SOLVE FOR Q THE REORDER QUANTITY
10    B=999999
20    FOR Q=1 TO 10000
30    C=.125*Q*(1-10000./125000.)/2+25*10000/Q
40    IF C < B THEN 60
50    GO TO 80
60    A1=Q
70    B=C
80    NEXT Q
90    PRINT A1,B
100   STOP
110   END

2085                 239.792
```

Therefore, 2085 units should be ordered each time to minimize the cost.

EXAMPLE 4.4

The Rapid River Company has finite replenishment equations $c_1 = .17Q1(1 - 900/5000)/2 + (37)900/Q1$ and $c_2 = .15Q2(1 - 600/4000)/2 + (29)600/Q2$ for products A and B, respectively. However, an additional problem is that each unit of A and B take up 1 cubic meter of storage space and the warehouse capacity is 1000 cubic meters. Therefore, the problem is to minimize $c = c_1 + c_2$ subject to $Q_1 \geqslant 0$, $Q_2 \geqslant 0$, and $Q_1 + Q_2 \leqslant 1000$. The program is given below (note that lines 20, 30, and 40 guarantee that Q_1 and Q_2 are less than or equal to 1000):

```
5     REM TWO VARIABLE (TWO PRODUCT)
6     REM COST MINIMIZATION PROBLEM
7     REM UNDER WAREHOUSE CONSTRAINTS
8     REM SOLVE FOR Q1 AND Q2
9     REM (THE REORDER QUANTITIES)
10    B=999999
20    FOR Q1=1 TO 1000
30    K=1000-Q1
40    FOR Q2=1 TO K
50    C1=.17*Q1*(1-900./5000.)/2.+37*900/Q1
60    C2=.15*Q2*(1-600./4000.)/2.+29*600/Q2
70    C=C1+C2
80    IF C < B THEN 100
90    GO TO 150
100   A1=Q1
110   B1=C1
120   A2=Q2
130   B2=C2
140   B=C
150   NEXT Q2
160   NEXT Q1
170   PRINT A1,B1
180   PRINT A2,B2
190   PRINT B
200   STOP
210   END

573               98.0533
427               67.9707
166.024
```

Therefore, Q_1 = 573 per order of Product A and Q_2 = 427 per order of Product B should minimize the inventory cost subject to the 1000 cubic meter capacity constraint.

It seems as though most realistic inventory models would be functions of three or four products and maybe a half dozen other variables or constraints. Using the Monte Carlo approach on these big models (trillions of feasible solutions) could lead to the statement and solution of very accurate inventory optimization problems.

Econometric Models

Much forecasting, predictive, and analytical work is done to set up a realistic econometric model of some type with profit, job satisfaction or gross domestic product, etc., as the response variable (a function of several independent variables). This is all well and good; however, once this is done, one should optimize the model by finding the correct settings for the variables to achieve the optimal response.

EXAMPLE 4.5

A forecasting study has revealed that the profit for products A and B, can be modeled as a function of advertising expenditures on the two products. The model is maximize $P = x^2 + 50x + 2800 - y^2 + 80y + 3200$ subject to $x \geqslant 0, y \geqslant 0$, and $x + y \leqslant 35$. Everything is in hundreds of dollars: $x + y \leqslant 35$ is the budget constraint (only \$3500 is available for advertising); x is the amount spent advertising Product A, while y is the amount spent advertising Product B. The program is below:

```
5     REM TWO VARIABLE (TWO PRODUCT)
6     REM SALES MAXIMIZATION BY
7     REM INCREASING ADVERTISING
8     REM SUBJECT TO A BUDGETARY CONSTRAINT
10    B=-999999
20    FOR X=0 TO 35
30    FOR Y=0 TO 35
40    IF X+Y > 35 THEN 130
50    P1=X**2+50*X+2800
60    P2=-Y**2+80*Y+3200
70    P=P1+P2
80    IF P > B THEN 100
90    GO TO 130
100   A1=X
110   A2=Y
120   B=P
130   NEXT Y
140   NEXT X
150   PRINT A1,A2
160   PRINT B
170   STOP
180   END

18          17
7358
```

Therefore, \$1800 advertising for A and \$1700 advertising for B should maximize the profit (other factors being held constant).

EXAMPLE 4.6

Suppose that your firm's economists have developed a productivity model, given as $P = 9826 - x_1{}^2 + 20x_1 - x_2{}^2 + 10x_2 - x_3{}^2 + 14x_3 + .14x_1x_2 + .15x_1x_3 + .19x_2x_3$ subject to $0 \leqslant x_1 \leqslant 25$, $0 \leqslant x_2 \leqslant 15$, and $0 \leqslant x_3 \leqslant 15$, where x_1 is the percent of budget spent on research two years ago; x_2 is the percent invested in capital equipment two years ago; x_3 is the percent invested in capital equipment five years ago; P is the resultant productivity index. Note that management has decreed that under no circumstances will research spending exceed 25% or will capital equipment expenditures exceed 15%.

Let us write a BASIC program to find the optimal solution to this model. The program which follows here is, of course, similar in structure to all of our other ones.

```
5     REM PRODUCTIVITY LEVEL
10    B=-999999
20    FOR X1=0 TO 25
30    FOR X2=0 TO 15
40    FOR X3=0 TO 15
50    P1=9826-X1**2+20*X1-X2**2+10*X2
60    P2=-X3**2+14*X3+.14*X1*X2
70    P3=.15*X1*X3+.19*X2*X3
80    P=P1+P2+P3
90    IF P > B THEN 110
100   GO TO 150
110   A1=X1
120   A2=X2
130   A3=X3
140   B=P
150   NEXT X3
160   NEXT X2
170   NEXT X1
180   PRINT A1,A2,A3
190   PRINT B
200   STOP
210   END

11              7              8
10028.6
```

Therefore, 11% spent on research two years ago, 7% invested in capital equipment two years ago, and 8% invested in capital equipment five years ago maximized productivity.

The company would also like to know if there are any nearly optimal answers (other possible courses of action). So we run the problem again and arrange to print any answers where the productivity index is greater than 10,027.5.

```
5     REM PRODUCTIVITY LEVEL
10    B=10027.5
20    FOR X1=0 TO 25
30    FOR X2=0 TO 15
40    FOR X3=0 TO 15
50    P1=9826-X1**2+20*X1-X2**2+10*X2
60    P2=-X3**2+14*X3+.14*X1*X2
70    P3=.15*X1*X3+.19*X2*X3
80    P=P1+P2+P3
90    IF P > B THEN 110
100   GO TO 150
110   PRINT X1,X2,X3
120   PRINT P
150   NEXT X3
160   NEXT X2
170   NEXT X1
200   STOP
210   END

10            6           8
10027.5
11            6           8
10028.6
11            6           9
10028.3
11            7           8
10028.6
11            7           9
10028.6
12            6           8
10027.6
12            6           9
10027.5
12            7           8
10027.8
12            7           9
10027.9
```

Therefore, we have several possible nearly optimal solutions.

Now let us assume that three years ago the firm spent 16% on capital equipment. Therefore, what should be the spending on capital equipment and research this year in order to maximize productivity two years hence? The program follows:

```
5     REM PRODUCTIVITY MODEL
10    B=-999999
20    FOR X1=0 TO 25
30    FOR X2=0 TO 15
40    X3=16
50    P1=9826-X1**2+20**X1-X2**2+10*X2
60    P2=-X3**2+14*X3+.14*X1*X2
70    P3=.15*X1*X3+.19*X2*X3
80    P=P1+P2+P3
90    IF P > B THEN 110
100   GO TO 150
110   A1=X1
120   A2=X2
130   A3=X3
140   B=P
160   NEXT X2
170   NEXT X1
180   PRINT A1,A2,A3
190   PRINT B
200   STOP
210   END

12          7             16
9972.84
```

Therefore, 12% should be spent on research and 7% on capital equipment to maximize productivity in the future.

EXAMPLE 4.7

The government is concerned about formation of investment capital. They feel that without sufficient capital to finance new private business ventures the country will have more economic problems. Therefore, government economists have experimented with a number of investment capital models. One such model is $C = 15{,}000 + x_1{}^{1.6} - .9x_2{}^2 + 11x_2 - x_3{}^2 + 17.4x_3 - .8x_4{}^2 + 16.2x_4 + 3.2x_3x_4 + 1.1x_1x_4$ subject to $3 \leqslant x_2 \leqslant 10$, $7 \leqslant x_3 \leqslant 15$, and $7 \leqslant x_4 \leqslant 14$, where x_1 is the percent increase in income over the 1970 base year; x_2 is the percent of income saved per household; x_3 is the prime interest rate percentage increase; x_4 is the treasury note rate percentage increase. The response variable C is available investment capital in tens of millions of dollars.

Given that x_1 is 18 and savings is projected to be 6%, maximize C. The program follows:

```
5    REM INVESTMENT CAPITAL MODEL
10   B=-999999
20   X1=18
30   X2=6
40   FOR X3=7 TO 15 STEP .1
50   FOR X4=7 TO 14 STEP .1
60   C1=15000+X1**1.6-.9*X2**2
70   C2=11*X2-X3**2+17.4*X3-.8*X4**2
80   C3=16.2*X4+3.2*X3*X4+1.1*X1*X4
90   C=C1+C2+C3
100  IF C > B THEN 120
110  GO TO 170
120  A1=X1
130  A2=X2
140  A3=X3
150  A4=X4
160  B=P
170  NEXT X4
180  NEXT X3
190  PRINT A1,A2,A3,A4
200  PRINT B
210  STOP
220  END

18          6          15.0000          14.0000
9752.10
```

Therefore, a 15% increase in the prime interest rate and a 14% increase in the treasury note rate should maximize investment capital.

EXAMPLE 4.8

The Ironwood Calculator Company has made a survey of their industry to try to determine the right combination of engineers and salespeople for their organization. One of the mathematical models developed from this study is $P = 4950 - 4x_1{}^2 + 56x_1 - x_2{}^2 + 24x_2 + 2.18x_1x_2$ subject to $4 \leqslant x_1 \leqslant 20$ and $5 \leqslant x_2 \leqslant 30$, where x_1 is the percentage of the work force that are salespeople; x_2 is the percentage of the work force that are engineers; P is the profit in thousands of German marks.

Note that management has stated that the percentage of salespeople and engineers will be between 4 and 20 and 5 and 30, respectively. The program is given on the following page.

```
5     REM ANALYSIS OF
10    REM THE WORKFORCE
20    B=-999999
30    FOR X1=4 TO 20
40    FOR X2=5 TO 30
50    P1=4950-4*X1**2+56*X1
60    P2=-X2**2+24*X2+2.18*X1*X2
70    P=P1+P2
80    IF P > B THEN 100
90    GO TO 130
100   A1=X1
110   A2=X2
120   B=P
130   NEXT X2
140   NEXT X1
150   PRINT A1,A2
160   PRINT B
170   STOP
180   END

15                    28
5693.60
```

Therefore, the company should have 15% salespeople and 28% engineers on its work force which could lead to a profit of 5693.6 thousands of marks (other things being equal).

EXAMPLE 4.9

The Chatham Steel Corporation is looking into the relationship of wages to profit for the next union bargaining session. They are also interested from a purely management point of view because they have a theory that their managers may be a little underpaid. From a study of wages and salaries in the steel industry, they have developed the following model: $P = 1100 - x_1{}^2 + 30x_1 - x_2{}^2 + 15.40x_2 + 1.2x_1x_2$ subject to $12 \leqslant x_1 \leqslant 20$ and $5 \leqslant x_2 \leqslant 10$, where x_1 is the average hourly wage paid to management (salary divided by 2000 work hours per year); x_2 is the average hourly wage paid to the workers; P is a profitability index.

Currently, Chatham Steel is paying management an average of $12.50 per hour and workers $7.50 per hour. Therefore, let us maximize the profitability function to see if adjustments could be recommended. The program is on the following page.

```
5     REM PROFIT AS A FUNCTION
6     REM OF MANAGEMENTS PAY
7     REM AND LABORS PAY
10    B=-999999
20    FOR X1=12 TO 20 STEP .1
30    FOR X2=5 TO 10 STEP .1
40    P1=1100-X1**2+30*X1
50    P2=-X2**2+15.40*X2+1.2*X1*X2
60    P=P1+P2
70    IF P > B THEN 90
80    GO TO 120
90    A1=X1
100   A2=X2
110   B=P
120   NEXT X2
130   NEXT X1
140   PRINT A1,A2
150   PRINT B
160   STOP
170   END

15.0000        7.70000
1384.40
```

Therefore, the company should pay management about \$15 per hour and labor about \$7.70 in order to maximize the profitability index. Of course, printouts of some of the nearly optimal solutions could shed light on the range of possibly good decisions.

EXAMPLE 4.10

The North Bay Glass Company has surveyed their industry and obtained the following profit model: $P = 100{,}900 - 3.8x_1{}^2 + 101x_1 - x_2{}^2 + 18x_2 - 3.9x_3{}^2 + 49x_3 - x_4{}^2 + 30x_4 + 1.5x_2x_4$ subject to $5 \leqslant x_i \leqslant 20$ for all i by management decree, where x_1 is the percent of the budget spent on management; x_2 is the percent spent on research; x_3 is the percent devoted to marketing and advertising; x_4 is the percent invested in capital equipment. The program to maximize P is as follows:

```
2     REM PROFIT AS A FUNCTION
3     REM SPENDING ON MANAGEMENT,
4     REM RESEARCH, MARKETING AND
5     REM CAPITAL EQUIPMENT
10    B=-999999
```

```
20    FOR X1=5 TO 20
30    FOR X2=5 TO 20
40    FOR X3=5 TO 20
50    FOR X4=5 TO 20
60    P1=-3.8*X1**2+101*X1
70    P2=-X2**2+18*X2
80    P3=-3.9*X3**2+49*X3
90    P4=-X4**2+30*X4
100   P5=100900+1.5*X2*X4
110   P=P1+P2+P3+P4+P5
120   IF P > B THEN 140
130   GO TO 190
140   A1=X1
150   A2=X2
160   A3=X3
170   A4=X4
180   B=P
190   NEXT X4
200   NEXT X3
210   NEXT X2
220   NEXT X1
230   PRINT A1,A2,A3,A4
240   PRINT B
250   STOP
260   END

13            20            6            20
102484
```

Therefore, 13% should be spent on management, 20% on research, 6% on marketing and advertising, and 20% of the budget should be spent on capital equipment.

However, the young executive who put this model together has run into a problem—the president of the company will not allow more than 20% of the budget to be spent on research and capital equipment. Therefore, the program is run over again with that constraint added in line 55, 55 IF X2+X4 > 20 THEN 190:

```
2     REM PROFIT AS A FUNCTION
3     REM OF SPENDING ON MANAGEMENT,
4     REM RESEARCH, MARKETING AND
5     REM CAPITAL EQUIPMENT
10    B=-999999
20    FOR X1=5 TO 20
```

```
30    FOR X2=5 TO 20
40    FOR X3=5 TO 20
50    FOR X4=5 TO 20
55    IF X2+X4 > 20 THEN 190
60    P1=-3.8*X1**2+101*X1
70    P2=-X2**2+18*X2
80    P3=-3.9*X3**2+49*X3
90    P4=-X4**2+30*X4
100   P5= 100900+1.5*X2*X4
110   P=P1+P2+P3+P4+P5
120   IF P > B THEN 140
130   GO TO 190
140   A1=X1
150   A2=X2
160   A3=X3
170   A4=X4
180   B=P
190   NEXT X4
200   NEXT X3
210   NEXT X2
220   NEXT X1
230   PRINT A1,A2,A3,A4
240   PRINT B
250   STOP
260   END

13        8        6        12
102164
```

Now the best answer has 8% for research and 12% spent on capital equipment.

EXAMPLE 4.11

An industrial psychologist has developed a motivation response model for a specific task. (Motivation is a function of perceived level of difficulty of the task.) The function is $P = -x^2 + 104x + 3000$. Find the level of difficulty x (on a scale of 0 to 200) that will maximize the motivation of the worker.

```
5    REM ONE VARIABLE
6    REM MAXIMIZATION
7    REM BY INCREMENTS
8    REM OF ONE TENTH
```

```
10    B=-999999
20    FOR X=0 TO 200 STEP .1
30    P=-X**2+104*X+3000
40    IF P > B THEN 60
50    GO TO 80
60    A1=X
70    B=P
80    NEXT X
90    PRINT A1,B
100   STOP
110   END

52.0000          5704
```

The program shows that it might be a good idea to adjust tasks so that their perceived level of difficulty is about 52.

EXERCISES

4.1 The Keweenaw Midland Corporation needs 15,000 units of redwood per year. The holding cost is $5 per unit per year. The ordering cost is $49 per order. Find the minimum cost order quantity solution using the basic EOQ model.

4.2 Keweenaw Midland also needs 22,000 units of Sitka spruce per year. The holding cost is $4 per unit per year. The ordering cost is $66 per order. Find the minimum cost order quantity solution using the basic EOQ model.

4.3 Keweenaw Midland wants to store the redwood and Sitka spruce in the same small warehouse, which has a capacity of 1000 units of wood (redwood and spruce units are the same size). Combine exercises 4.1 and 4.2 with the constraint $Q_1 + Q_2 \leqslant 1000$ and find this minimum cost solution.

4.4 The Brookville Hoover Corporation buys units of industrial ceramics from Beaconsfield Voss according to the following price list:

Number of units	Unit price ($)
0 - 19,999	3.50
20,000 - 99,999	3.00
100,000 - 299,999	2.80
300,000 - 999,999	2.60
1,000,000 or over	2.50

The unit holding cost per year is \$.22. The ordering cost is \$50 per shipment. The revenue per unit sold by Brookville Hoover is estimated at:

Number of units	Unit price (\$)
0 - 49,999	15.00
50,000 - 249,999	13.00
250,000 - 749,999	12.00
750,000 - 1,999,999	10.00
2,000,000 or more	9.50

Brookville will finance this deal by borrowing up to \$100,000 at 10%, the next \$200,000 at 11%, and everything over \$300,000 is borrowed at 14%. Find the best inventory policy for Brookville.

4.5 The Cooperstown Co. makes shaping glass and decorating glass. They have an inventory of 5000 units of shaping glass and 7500 units of decorating glass. Management wants 20,000 units of each in inventory at the end of the coming year. The price-quantity curve for shaping glass is $P_1 = -.000002x_1 + 31$, and $P_2 = -.000003x_2 + 35$ for decorating glass. Fixed advertising costs are \$2000 for shaping glass and \$4000 for decorating glass for this year. Both types of glass cost \$4 per unit to make up to 50,000 total, and \$6 per unit (due to overtime) for every unit over 50,000.

The company will borrow money at 11% to purchase the ingredients to make the glass. The price list is below:

Shaping glass ingredients per unit	Unit cost (\$)	Decorating glass ingredients per unit	Unit cost (\$)
0 - 19,999	1.95	0 to 24,999	2.25
20,000 - 49,999	1.80	25,000 to 99,999	2.00
50,000 - 229,999	1.70	100,000 to 249,000	1.65
230,000 or more	1.60	Over 249,000	1.30

Find the maximum profit solution.

4.6 A study of the brickmaking industry indicates that productivity is a function of pay level (x_1) and perceived level of authority of their boss (x_2).

The forecasting model is $P = 18{,}000 - 2x_1^2 + 172x_1 - x_2^2 + 117x_2 + .003x_1x_2$ subject to $0 \leqslant x_1 \leqslant 100$ and $0 \leqslant x_2 \leqslant 100$, where x_1 is the worker wage per day and x_2 is the authority rating of their boss on a 0 to 100 scale; 0 would mean no authority, 100 very domineering, and 50 about average. Find the maximum productivity combination.

4.7 Sales of records has been modeled as a function of price of record (x_1), popularity of artist (x_2), advertising expenditures (x_3), and quality of the song (x_4). The model function is: $S = 210{,}000 - 3x_1^2 + 1292x_1 + 500x_2$

$+ 600x_3^{.94} + 5600x_4 + 912x_2x_4 + 29x_1x_3$ subject to $0 \leqslant x_1 \leqslant 1000$ cents, $0 \leqslant x_2 \leqslant 100, 0 \leqslant x_3 \leqslant 50{,}000$, and $0 \leqslant x_4 \leqslant 100$.

Given that a company can spent \$10,000 on advertising and the artist is rated at 75 and the record is rated at 90, what is the optimal price (maximizing sales) to charge?

4.8 Job satisfaction is modeled as a function of perceived working conditions by the employee on a scale of 0 to 100, 100 being the perception of the best possible conditions. The model is $S = 170 - x^2 + 164x$ for $0 \leqslant x \leqslant 100$.

Find the level of perception of working conditions associated with maxmum job satisfaction.

4.9 Wall Street analysts rated companies as to general "health" of the company versus amount of government regulation affecting them. 0 is no regulation, 100 excessive regulation, and 50 about average. The model arrived at is $P = 200 - 3x_1^2 + 50x_1$ subject to $0 \leqslant x_1 \leqslant 100$.

Find the level of regulation that seems best for industry.

4.10 A survey of ten industrialized countries asked businesspeople to indicate how easy it was to get capital for their businesses. These were compared with the general level of taxation of the average citizen of that country. The subsequent model is $P = 122 - x^2 + 46x$ subject to $10 \leqslant x \leqslant 60$, where x is the percent of income paid in taxes.

Find the optimal tax level to stimulate private investment.

4.11 The health of the Camelot economy has been rated for the last thirty years and the variable x equals the percentage of the GNP that the government takes in, in taxes. The model is $H = 100 - x^2 - 6x$ for $0 \leqslant x \leqslant 40$.

Find x that is best for the health of the country.

4.12 Profit of a company is modeled as a function of number of workers versus number of machines (\$20,000 capital investment equals one machine). $P = 1000 - 2.8x_1^2 + 182x_1 - 3.1x_2^2 + 309x_2 + .025x_1x_2$ subject to $0 \leqslant x_1 \leqslant 500$ workers and $0 \leqslant x_2 \leqslant 800$ units of \$20,000 for machines. Maximize P.

4.13 The Okemos Corporation is studying productivity in their industry as a function of the rate of pay to hourly workers and the dollar amount of stock the average hourly worker owns in the corporation. The resulting model is $P = 1502 - 9x_1^2 + 90x_1 + .65x_2 + .04x_1x_2$ subject to $0 \leqslant x_1 \leqslant \$12$ and $0 \leqslant x_2 \leqslant \$5000$, where x_1 is the hourly wage and x_2 is the average dollar amount of the stock in the company that the worker owns.

To illuminate these free enterprise forces, maximize P.

4.14 Maximize $L = 8 - .9x^2 + 6x$, where L is the demand for labor in an average small company in a particular industry and x is the minimum wage ($\$2 \leqslant x \leqslant \$4$).

Suggested Reading

Archibald, G. and Lipsey, R. *An Introduction to Mathematical Economics Method and Application*. New York: Harper & Row, 1976.

Johnston, J. *Econometric Methods*, 2nd ed. New York: McGraw-Hill, 1972.

McMillan, C. Gonzalez. *Systems Analysis, A Computer Approach to Decision Models*. Homewood, IL.: Richard D. Irwin, 1973.

CHAPTER 5

Chemical Yield Equations

Chemical yield equations are of considerable importance to the chemical industry. Almost every chemical experiment or process is designed to *maximize* the yield of a chemical or compound or *minimize* the amount of a pollutant released or *minimize* the cost of the process.

The most efficient way to find the best course of action in these problems is to try different combinations of additives and/or vary the temperatures and pressure (as appropriate) and record the resulting yield of the compound in question. Then a "least squares" best polynomial response curve is fitted to the data. At that point, the programs in this chapter (and chapter 12) can be used to optimize the function.

EXAMPLE 5.1

A chemical process has been run a number of times with different combinations of two additives. From this, a nonlinear predictive mathematical model of the yield of the process was developed. It is $y = 3850 - x_1{}^2 + 100x_1 - x_2{}^2 + 60x_2 + .1x_1x_2$. We seek to find the amounts x_1 and x_2 of additives A and B that will maximize the yield. From the chemical theory, we know that additive A cannot exceed 60 units and additive B cannot exceed 65 units. We want to maximize y subject to $0 \leqslant x_1 \leqslant 60$ and $0 \leqslant x_2 \leqslant 65$. The program follows:

```
5    REM CHEMICAL YIELD
6    REM EQUATION OPTIMIZATION
10   B=-999999
```

```
20    FOR X1=0 TO 60
30    FOR X2=0 TO 65
40    Y=3850-X1**2+100*X1-X2**2+60*X2+.1*X1*X2
50    IF Y > B THEN 70
60    GO TO 100
70    B=Y
80    A1=X1
90    A2=X2
100   NEXT X2
110   NEXT X1
120   PRINT A1,A2
130   PRINT B
140   STOP
150   END

52              33
7408.6
```

Therefore, 52 units of A and 33 units of B should give the optimal yield of 7408.6. Also, if necessary, one could rerun the program to get the optimal solution to the nearest tenths:

```
5     REM CHEMICAL YIELD
6     REM EQUATION OPTIMIZATION
10    B=-999999
20    FOR X1=0 TO 60 STEP .1
30    FOR X2=0 TO 65 STEP .1
40    Y=3850-X1**2+100*X1-X2**2+60*X2+.1*X1*X2
50    IF Y > B THEN 70
60    GO TO 100
70    B=Y
80    A1=X1
90    A2=X2
100   NEXT X2
110   NEXT X1
120   PRINT A1,A2
130   PRINT B
140   STOP
150   END

51.6000               32.6000
7408.896
```

This type of accuracy is rarely justified in a chemistry experiment due to experimental and forecasting error, etc. However, in cases where it is justified, it can be obtained.

Now let us say that we have the same problem as above, but that if x_1 and x_2 combined exceeds 70 then the reaction might explode. So let us run the program over again to maximize y subject to $0 \leqslant x_1 \leqslant 60$, $0 \leqslant x_2 \leqslant 65$, and $x_1 + x_2 \leqslant 70$:

```
5    REM CHEMICAL YIELD
6    REM EQUATION OPTIMIZATION
7    REM UNDER CONSTRAINTS
10   B=-999999
20   FOR X1=0 TO 60 STEP .1
30   FOR X2=0 TO 65 STEP .1
35   IF X1+X2 > 70 THEN 100
40   Y=3850-X1**2+100*X1-X2**2+60*X2+.1*X1*X2
50   IF Y > B THEN 70
60   GO TO 100
70   B=Y
80   A1=X1
90   A2=X2
100  NEXT X2
110  NEXT X1
120  PRINT A1,A2
130  PRINT B
140  STOP
150  END

44.5000           25.5000
7312.975
```

Therefore, 44.5 units of A and 25.5 units of B should be added.

EXAMPLE 5.2

A chemical yield process has been modeled as $P = 5000 + 2x_1 + 5.2x_2 + 3.8x_3 + .92x_1x_2 - .0075x_1x_3 + 3.2x_2x_3 + .02x_1x_2x_3$ subject to $80 \leqslant x_1 \leqslant 200$, $0 \leqslant x_2 \leqslant 10$, and $0 \leqslant x_3 \leqslant 50$, where x_1 is the temperature, x_2 is the pressure, and x_3 is the length of reaction time. We would like to maximize P. The program is:

```
5    REM CHEMICAL YIELD AS A
6    REM FUNCTION OF TEMP,
7    REM PRESSURE AND LENGTH
8    REM OF REACTION TIME.
10   B=-999999
20   FOR X1=80 TO 200
30   FOR X2=0 TO 10
```

```
40    FOR X3=0 TO 50
45    IF X1+X3 > 215 THEN 200
50    P1=5000+2*X1+5.2*X2+3.8*X3
60    P2=.92*X1*X2-.0075*X1*X3+3.2*X2*X3
70    P3=.02*X1*X2*X3
75    P=P1+P2+P3
80    IF P > B THEN 100
90    GO TO 200
100   A1=X1
110   A2=X2
120   A3=X3
130   B=P
200   NEXT X3
210   NEXT X2
220   NEXT X1
230   PRINT A1,A2,A3
240   PRINT B
250   STOP
260   END

165         10            50
10278.1
```

Therefore, 165 degrees and 10 atmospheres of pressure for 50 seconds will maximize the yield.

EXAMPLE 5.3

Let us maximize $P = 68{,}100 - x_1{}^2 + 260x_1 - x_2{}^2 + 400x_2 + .06x_1x_2$ subject to $100 \leqslant x_1 \leqslant 250$ and $0 \leqslant x_2 \leqslant 250$, where x_1 is the temperature in Celsius and x_2 is the reaction time in seconds.

```
5     REM CHEMICAL YIELD AS
6     REM A FUNCTION OF TEMPERATURE
7     REM AND REACTION TIME
10    B=-999999
20    FOR X1=100 TO 250
30    FOR X2=0 TO 250
40    P1=68100-X1**2+260*X1
50    P2=-X2**2+400*X2+.06*X1*X2
60    P=P1+P2
70    IF P > B THEN 90
80    GO TO 120
90    A1=X1
100   A2=X2
110   B=P
```

```
120  NEXT X2
130  NEXT X1
135  PRINT A1,A2
138  PRINT B
140  STOP
150  END

136            204
126613
```

Therefore, 136 degrees and 204 seconds should yield a maximum of 126,613 units.

However, let us assume in the above problem that it is not possible to maintain the temperature above 115 degrees for more than 3 minutes. So we modify the program (lines 32, 33, and 35) to allow for this, as follows:

```
5    REM CHEMICAL YIELD AS
6    REM A FUNCTION OF TEMPERATURE
7    REM AND REACTION TIME
8    REM BUT CANT MAINTAIN TEMP ABOVE 115
9    REM FOR MORE THAN 3 MINUTES
10   B=-999999
20   FOR X1=100 TO 250
30   FOR X2=0 TO 250
32   IF X1 < 116 THEN 35
33   GO TO 120
35   IF X2 > 180 THEN 120
40   P1=68100-X1**2+260*X1
50   P2=-X2**2+400*X2+.06*X1*X2
60   P=P1+P2
70   IF P > B THEN 90
80   GO TO 120
90   A1=X1
100  A2=X2
110  B=P
120  NEXT X2
130  NEXT X1
135  PRINT A1,A2
138  PRINT B
140  STOP
150  END

115          180
125617
```

So 115 degrees maintained for 180 seconds yields a maximum of $P = 125{,}617$.

EXAMPLE 5.4

A chemical model designed to increase the purity of high-grade paper by the introduction of various amounts of three additives is developed. It is $P = 16{,}525 + .009x_2x_3 - 4x_1^2 + 120x_1 - 9x_2^2 + 180x_2 - x_3^2 + 90x_3 + .007x_1x_2 - .004x_1x_3$ subject to $0 \leqslant x_1 \leqslant 50$, $0 \leqslant x_2 \leqslant 50$, and $40 \leqslant x_3 \leqslant 50$. Let us write a program to maximize this function:

```
5    REM CHEMICAL PURITY OF HIGH GRADE PAPER
6    REM AS A FUNCTION OF THREE ADDITIVES X1, X2, X3
9    B=-999999
10   FOR X1=0 TO 50
20   FOR X2=0 TO 50
30   FOR X3=40 TO 50
40   P1=16525+.009*X2*X3
50   P2=-4*X1**2+120*X1
60   P3=-9*X2**2+180*X2
70   P4=-X3**2+90*X3
80   P5=.007*X1*X2-.004*X1*X3
90   P=P1+P2+P3+P4+P5
100  IF P > B THEN 120
110  GO TO 160
120  A1=X1
130  A2=X2
140  A3=X3
150  B=P
160  NEXT X3
170  NEXT X2
180  NEXT X1
190  PRINT A1,A2,A3
200  PRINT B
210  STOP
220  END

15            10            45
20352.4
```

Therefore, 15 units of additive 1, 10 units of additive 2, and 45 units of additive 3 yield a maximum purity index of 20,352.4.

Now suppose that if $x_1 + x_2$ is greater than 20 then there is too much streaking in the paper. We rerun the program with that condition in line 35:

```
5     REM CHEMICAL PURITY OF HIGH GRADE PAPER
6     REM AS A FUNCTION OF THREE ADDITIVES X1,X2,X3
7     REM WITH CONSTRAINT TO PREVENT STREAKING
9     B=-999999
10    FOR X1=0 TO 50
20    FOR X2=0 TO 50
30    FOR X3=40 TO 50
35    IF X1+X2 > 20 THEN 160
40    P1=16525+.009*X2*X3
50    P2=-4*X1**2+120*X1
60    P3=-9*X2**2+180*X2
70    P4=-X3**2+90*X3
80    P5=.007*X1*X2-.004*X1*X3
90    P=P1+P2+P3+P4+P5
100   IF P > B THEN 120
110   GO TO 160
120   A1=X1
130   A2=X2
140   A3=X3
150   B=P
160   NEXT X3
170   NEXT X2
180   NEXT X1
190   PRINT A1,A2,A3
200   PRINT B
210   STOP
220   END

12       8            45
20279.8
```

Therefore, 12 units of additive 1, 8 units of additive 2, and 45 units of additive 3 will maximize the purity index subject to $x_1 + x_2 \leqslant 20$.

EXAMPLE 5.5

Let us maximize $P = 401{,}332{,}244 - |x_1 - 110|^3 - x_2{}^2 + 44x_2 - 484 - |x_3 - 12|^3 + .00008x_1x_2{}^2x_3$ subject to $80 \leqslant x_1 \leqslant 150$, $10 \leqslant x_2 \leqslant 30$, and $0 \leqslant x_3 \leqslant 20$, where x_1 is the temperature in Celsius, x_2 is the pressure in atmospheres, and x_3 is the reaction time in seconds.

```
5     REM CHEMICAL YIELD AS
6     REM A FUNCTION OF TEMPERATURE,
7     REM PRESSURE AND REACTION TIME
10    B=-999999
20    FOR X1=80 TO 150
```

```
30    FOR X2=10 TO 30
40    FOR X3=0 TO 20
50    P1=401332244
60    P2=ABS((X1-110)**3)
70    P3=-X2**2+44*X2-484
80    P4=ABS((X3-12)**3)
90    P5=.00008*X1*X2**2*X3
100   P=P1-P2+P3-P4+P5
110   IF P > B THEN 130
120   GO TO 170
130   A1=X1
140   A2=X2
150   A3=X3
160   B=P
170   NEXT X3
180   NEXT X2
190   NEXT X1
200   PRINT A1,A2,A3
210   PRINT B
220   STOP
230   END

110               25              13
4.01332E+08
```

Therefore, x_1 = 110 degrees, x_2 = 25 atmospheres, and x_3 = 13 seconds yields a maximum of 401,332,000 units.

Now suppose that researchers are worried that if $2x_1 + 3x_2$ is greater than 200 then an explosion could occur. So we run the program over again with this constraint in line 45.

```
5     REM CHEMICAL YIELD AS
6     REM A FUNCTION OF TEMPERATURE,
7     REM PRESSURE AND REACTION TIME
8     BUT SUBJECT TO A CONSTRAINT
9     REM TO PREVENT AN EXPLOSION
10    B=-999999
20    FOR X1=80 TO 150
30    FOR X2=10 TO 30
40    FOR X3=0 TO 20
45    IF 2*X1+3*X2 > 200 THEN 170
50    P1=401332244
60    P2=ABS((X1-110)**3)
70    P3=-X2**2+44*X2-484
80    P4=ABS((X3-12)**3)
```

```
90    P5=.00008*X1*X2**2*X3
100   P=P1-P2+P3-P4+P5
110   IF P > B THEN 130
120   GO TO 170
130   A1=X1
140   A2=X2
150   A3=X3
160   B=P
170   NEXT X3
180   NEXT X2
190   NEXT X1
200   PRINT A1,A2,A3
210   PRINT B
220   STOP
230   END

85            10          12
4.01316E+08
```

Therefore, $x_1 = 85$ degrees, $x_2 = 10$ atmospheres, and $x_3 = 12$ seconds should maximize the yield.

EXAMPLE 5.6

The San Diego Chemical Company has developed a chemical process that is a function of six variables:

- x_1 pressure in atmospheres
- x_2 temperature in Celsius
- x_3 reaction time in seconds
- x_4 weight of catalyst 4 in grams
- x_5 volume of catalyst 5 in milliliters
- x_6 time at which catalyst 5 is added

The equation to be maximized is $P = -x_1^4 + 40x_1^3 - 600x_1^2 - 1.04x_2^2 + 253x_2 - .98x_3^2 + 200x_3 - .95x_4^2 + 19x_4 - .88x_5^2 + 81x_5 - x_6^2 + 100x_6 + 3{,}937{,}000 + 4000x_1 + .045x_1x_3 + .025x_4x_5 + .07x_2x_3 - .03x_3x_6$ subject to $5 \leqslant x_1 \leqslant 15$, $85 \leqslant x_2 \leqslant 150$, $60 \leqslant x_3 \leqslant 180$, $4 \leqslant x_4 \leqslant 20$, $0 \leqslant x_5 \leqslant 100$, and $0 \leqslant x_6 \leqslant x_3$. There are 11 x 66 x 121 x 17 x 101 = 150,831,582 combinations of the first five variables without even considering x_6 which would run the combinations into the trillions. Therefore, we will start with a Monte Carlo simulation and then focus in to find the true optimum.

```
5     REM CHEMICAL YIELD AS A FUNCTION
6     REM OF PRESSURE, TEMPERATURE, REACTION
7     REM TIME, WEIGHT OF CATALYST X4, VOLUME
8     REM OF CATALYST X5 AND THE TIME (X6)
9     REM AT WHICH YOU ADD CATALYST X5
10    X=1
12    B=-999999
20    FOR I=1 TO 10000
30    X1=5+INT(RND(X)*11)
40    X2=85+INT(RND(X)*66)
50    X3=60+INT(RND(X)*121)
60    X4=4+INT(RND(X)*17)
70    X5=INT(RND(X)*101)
80    X6=INT(RND(X)*(X3+1))
90    P1=-X1**4+40*X1**3-600*X1**2
100   P2=-1.04*X2**2+253*X2
110   P3=-.98*X3**2+200*X3
120   P4=-.95*X4**2+19*X4
130   P5=-.88*X5**2+81*X5
140   P6=-X6**2+100*X6
150   P7=3937000+4000*X1
160   P8=.045*X1*X3+.025*X4*X5
170   P9=.07*X2*X3-.03*X3*X6
180   P=P1+P2+P3+P4+P5+P6+P7+P8+P9
190   IF P > B THEN 210
200   GO TO 280
210   A1=X1
220   A2=X2
230   A3=X3
240   A4=X4
250   A5=X5
260   A6=X6
70    B=P
280   NEXT I
290   PRINT A1,A2,A3,A4,A5,A6
300   PRINT B
310   STOP
320   END
```

```
9              127            108            15             41
50
3.97780E+06

11             125            101            14             47
45
3.97781E+06

12             119            104            11             43
51
3.97779E+06
```

After a few printouts, we decide to limit x_3 to between 100 and 110 and x_5 to between 40 and 50.

```
  5     REM CHEMICAL YIELD AS A FUNCTION
  6     REM OF PRESSURE, TEMPERATURE, REACTION
  7     REM TIME, WEIGHT OF CATALYST X4, VOLUME
  8     REM OF CATALYST X5 AND THE TIME (X6)
  9     REM AT WHICH YOU ADD CATALYST X5
 10     X=1
 12     B=-999999
 20     FOR I=1 to 10000
 30     X1=5+INT(RND(X)*11)
 40     X2=85+INT(RND(X)*66)
 50     X3=100+INT(RND(X)*11)
 60     X4=4+INT(RND(X)*17)
 70     X5=40+INT(RND(X)*11)
 80     X6=INT(RND(X)*(X3+1))
 90     P1=-X1**4+40*X1**3-600*X1**2
100     P2=-1.04*X2**2+253*X2
110     P3=-.98*X3**2+200*X3
120     P4=-.95*X4**2+19*X4
130     P5=-.88*X5**2+81*X5
140     P6=-X6**2+100*X6
150     P7=3937000+4000*X1
160     P8=.045*X1*X3+.025*X4*X5
170     P9=.07*X2*X3-.03*X3*X6
180     P=P1+P2+P3+P4+P5+P6+P7+P8+P9
190     IF P > B THEN 210
200     GO TO 280
210     A1=X1
220     A2=X2
230     A3=X3
240     A4=X4
250     A5=X5
260     A6=X6
270     B=P
280     NEXT I
290     PRINT A1,A2,A3,A4,A5,A6
300     PRINT B
310     STOP
320     END
```

```
11      125       105       9        45
51
3.97785E+06

9       127       106       11       48
49
3.99784E+06
```

```
10          126            107            12              46
47
3.97785E+06
11          128            105            7               46
50
3.97783E+06
10          125            105            12              45
46
3.97784E+06
11          127            107            8               45
49
3.97784E+06
10          127            107            9               47
48
3.97785E+06
12          127            105            10              47
47
3.97784E+06
```

After seeing these printouts it is fairly obvious that the maximal solution is in the range $9 \leqslant x_1 \leqslant 12$, $125 \leqslant x_2 \leqslant 127$, $105 \leqslant x_3 \leqslant 107$, $7 \leqslant x_4 \leqslant 12$, $45 \leqslant x_5 \leqslant 48$, and $46 \leqslant x_6 \leqslant 51$. The program modified to look at all those prints is below:

```
5     REM CHEMICAL YIELD AS A FUNCTION
6     REM OF PRESSURE, TEMPERATURE, REACTION
7     REM TIME, WEIGHT OF CATALYST X4, VOLUME
8     REM OF CATALYST X5 AND THE TIME (X6)
9     REM AT WHICH YOU ADD CATALYST X5
10    X=1
12    B=-999999
20    FOR X1=9 TO 12
30    FOR X2=125 TO 127
40    FOR X3=105 TO 107
50    FOR X4=7 TO 12
60    FOR X5=45 TO 48
70    FOR X6=46 TO 51
90    P1=-X1**4+40*X1**3-600*X1**2
100   P2=-1.04*X2**2+253*X2
110   P3=-.98*X3**2+200*X3
120   P4=-.95*X4**2+19*X4
130   P5=-.88*X5**2+81*X5
```

```
140  P6=-X6**2+100*X6
150  P7=3937000+4000*X1
160  P8=.045*X1*X3+.025*X4*X5
170  P9=.07*X2*X3-.03*X3*X6
180  P=P1+P2+P3+P4+P5+P6+P7+P8+P9
190  IF P > B THEN 210
200  GO TO 280
210  A1=X1
220  A2=X2
230  A3=X3
240  A4=X4
250  A5=X5
260  A6=X6
270  B=P
280  NEXT X6
281  NEXT X5
282  NEXT X4
283  NEXT X3
284  NEXT X2
285  NEXT X1
290  PRINT A1,A2,A3,A4,A5,A6
300  PRINT B
310  STOP
320  END

11        125        106        11        46
48
3.97786E+06
```

Therefore

x_1 = 11 atmospheres
x_2 = 125 degrees Celsius
x_3 = 106 seconds reaction time
x_4 = 11 grams of catalyst 4
x_5 = 46 milliliters of catalyst 5
x_6 = 48 seconds into the reaction is the time to add catalyst 5
P is maximized at 3,977,860 units.

Another possible application would be to model a chemical process and take the costs of various ingredients and alternative chemicals into account (discounts for buying in quantity, etc.). Then the objective would be to maximize yield subject to various cost constraints or to minimize the cost of the process subject to certain quality and/or yield standards.

EXERCISES

5.1 Maximize $P = 1528 + 6x_1 + 2x_2 + 3x_3 - 2x_1x_2 + 1.6x_1x_3 - .9x_2x_3 + 2x_1x_2x_3$ subject to $x_1 \geqslant 0$, $x_2 \geqslant 0$, $x_3 \geqslant 0$, and $1.5x_1 + 2x_2 + x_3 \leqslant 1000$, where x_1 is the temperature in Celsius, x_2 is the pressure in atmospheres, and x_3 is the length of reaction time in seconds. Note: The chemists have estimated that if any combination of 1.5 times temperature plus 2 times pressure plus length of reaction exceeds 1000, the process will blow up, hence the constraint. *Hint:* Do a Monte Carlo optimization run or two and then focus in and search for and find the true optimum.

5.2 Maximize $Y = 1819 - 2x_1{}^2 + 108x_1 - 1.6x_2{}^2 + 91x_2 + x_1x_2$ subject to $0 \leqslant x_1 \leqslant 400$ and $0 \leqslant x_2 \leqslant 250$, where x_1 is the temperature and x_2 is the length of the reaction.

5.3 Maximize $Y = 2072 - 3.6x_1{}^2 + 216x_1$ subject to $0 \leqslant x_1 \leqslant 2000$, where x_1 is the length of the reaction in seconds.

5.4 Maximize $Y = 10{,}406 - 8x_1{}^2 + 29x_1 - 2x_2{}^2 + 103x_2 - 6.1x_3{}^2 + 105x_3 - .4x_4{}^2 + 202x_4 - 1.6x_5{}^2 + 75x_5 + x_1x_2x_3 - x_1x_4x_5 + 2x_4x_5$ for $0 \leqslant x_1 \leqslant 1000$, $0 \leqslant x_2 \leqslant 100$, $0 \leqslant x_3 \leqslant 1000$, $0 \leqslant x_4 \leqslant 50$, and $0 \leqslant x_5 \leqslant x_3$. x_1 is the temperature in Celsius; x_2 is the pressure in atmospheres; x_3 is the length of reaction time in seconds; x_4 is the number of units of Catalyst A; x_5 is when to add the number of units of Catalyst A.

5.5 Maximize $Y = 6204 - x_1{}^2 + 65x_1 - 2x_2{}^2 + 107x_2 - 5.8x_2{}^2 + 204x_3 + x_1x_3 - x_1x_2 + .002x_1x_2x_3$, where $0 \leqslant x_1 \leqslant 300$ (temperature), $0 \leqslant x_2 \leqslant 100$ units of Catalyst 1, $0 \leqslant x_3 \leqslant 100$ units of Catalyst 2, and $x_1 + x_2 \leqslant 163$ or the reaction becomes unstable.

5.6 The French River Fuel Company is planting fibrous Plant K on its fuel farms. It then harvests and processes Plant K and turns it into methanol gas that the company sells for \$6 per gas unit. Each acre has a \$50 planting, harvesting, and converting cost. However, after experimenting with increasing the production per acre by adding units of Fertilizer 1 (x_1 is the number of units per acre) and units of Nutrient 2 (x_2 is the number of units per acre), the yield per acre equation was modeled to be $Y = -500 - x_1{}^2 + 60x_1 - x_2{}^2 + 50x_2 + .02x_1x_2$, where Y is the yield of methanol gas per acre in gas units.

However, Fertilizer 1 costs \$18 per unit and Nutrient 2 costs \$22 per unit. What the company would like to do is to maximize the profit per acre, not just the yield. Note that x_1 and x_2 can be no larger than 200 units in total treatment (at that point the yield would surely drop, according to agricultural experts). So we have maximize $P = \$6(-500 - x_1{}^2 + 60x_1 - x_2{}^2 + 50x_2 + .02x_1x_2) - 50 - 18x_1 - 22x_2$ subject to $x_1 \geqslant 0$, $x_2 \geqslant 0$, and $x_1 + x_2 \geqslant 200$.

Write a BASIC program to solve this system.

5.7 If the price of Fertilizer 1 jumps to \$25 a unit and the price of Nutrient 2 jumps to \$35 per unit, what is the maximum profit solution for French River Fuel Company?

5.8 French River has an option to buy an extra one hundred acres of good farmland adjacent to its fuel farm. The price is \$250,000. They can finance it with a twenty-year mortgage at 10% interest (no money down because of government help for fuel farms). The land is estimated to appreciate in value at 7% a year for the next twenty years. It has the same growing conditions as their other land so they plan to use it to plant fibrous Plant K and turn it into methanol gas and sell it. Is it worth it to make the deal? Find the solution to maximize profit.

5.9 Solve exercise 5.8 again assuming that the price of methanol gas increases 5% a year for the next twenty years. Find the solution to maximize profit.

5.10 Hybrid G Plus corn is being tested at an experimental station in Iowa. It has passed all of the preliminary tests and is known to be a top-grade corn with good market potential. The question is what type of fertilizer should be used on it and in what quantity. Fertilizers A and B are known to be good separately, so the experimental farm tries them in combination. After 40 sample plantings with different treatment levels of A and B, the regression statistical model is determined to be $Y = -4800 - x^2 - y^2 + 80x + 120y - .075xy$ subject to $0 \leqslant x \leqslant 100$ and $0 \leqslant y \leqslant 100$, where x equals number of units A and y equals the number of units of B applied to the corn. Y is the resultant estimated bushels per acre yield. The bounds of 100 were recommended by experienced agriculture scientists. Find the combination of units of A and B that will maximize the forecasted yield.

5.11 The Albion Company has been searching for the right plant to grow in quantity cheaply to harvest and turn into methanol gas for fuel. After much work, they have come up with Plant X4 which grows so well that they believe it will be cost effective and can compete right away with conventional fuels. To help Plant X4 grow bigger and faster, combinations of three nutrients (1, 2, and 3) are being tried. The response curve is $Y = 108 + .2x_1 + .3x_2 + .15x_3 + .05x_1x_2 - .06x_1x_3 + .02x_2x_3$, and $x_1 + 1.2x_2 + .8x_3 \leqslant 70$ due to cost limitations if plant X4 is to be profitable. x_i is the number of units of nutrient i. Maximize Y.

Suggested Reading

Beech, G. *Fortran IV in Chemistry–An Introduction to Computer-Assisted Methods.* New York: John Wiley & Sons, 1975.

Nickerson, C. and Nickerson, I. *Statistical Analysis for Decision Making.* Princeton, N.J.: Petrocelli Books, 1978.

Skarabis, H. and Sint, P. *COMPSTAT 1–Lectures in Computational Statistics.* Wurzburg, Vienna: Physica Verlag 1978.

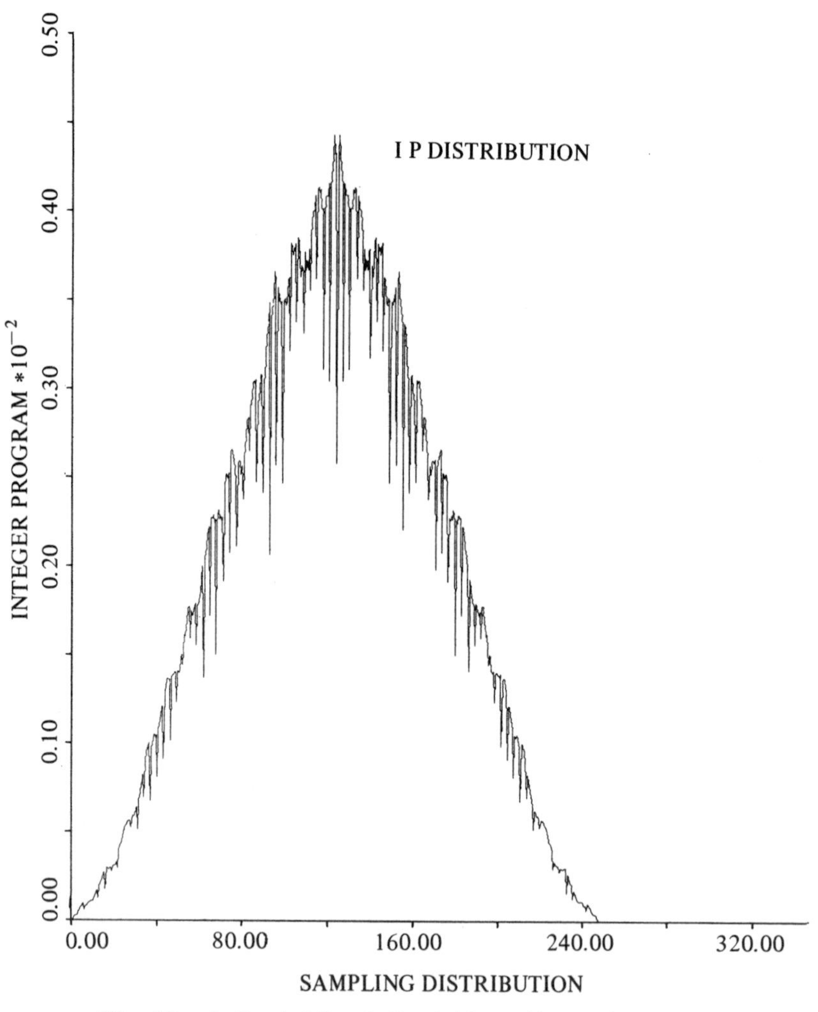

SAMPLING DISTRIBUTION

of $P = 10x_1 + .5x_2 + 3.8x_3 + .5x_4 + 10x_5$ subject to $0 \leqslant x_i \leqslant 10 \quad i = 1, 5$

FIGURE 6.1 is the distribution of all the answers of $P = 10x_1 + .5x_2 + 3.8x_3 + .5x_4 + 10x_5$ subject to $0 \leqslant x_i \leqslant 10, i = 5$. We do not need the distribution of all answers in order to find the true optimum. But by studying these we can develop the multistage Monte Carlo integer programs that help us to find the true optimum of large-scale problems. Examples are in chapter 12.

CHAPTER 6

Pharmacology and Medical Research

When medical researchers test combinations of drugs and other treatments to cure specific illnesses, they collect data on the performance of the various amounts of each substance in combination. Then a simple polynomial response curve is drawn from the data.

This chapter (and chapter 12) can be used to find the maximum or minimum of that function, as the usual goal is for the doctor to *minimize* the number of days until recovery or *minimize* pain or *maximize* the reduction in blood pressure or *maximize* the reduction in tumor size or *maximize* the concentration of a certain chemical in the blood stream.

EXAMPLE 6.1

Drugs A and B are known for their ability to cure Condition K. However, the treatment period is long and painful so doctors are beginning to experiment with giving combinations of A and B.

The response variable D is the number of days that it takes to cure the patient. A preliminary study has yielded the following response function: $D = 53 + x_1{}^2 - 8x_1 + x_2{}^2 - 12x_2 - .10x_1x_2$, where x_1 is the number of units of Drug A given to the patient per day and x_2 is the number of units of Drug B given to the patient per day, and $0 \leqslant x_1 \leqslant 7$ units, $0 \leqslant x_2 \leqslant 7$ units for both drugs because patients' systems cannot tolerate more than 7 units of each per day.

Below is the program to find the combination of units of drugs A and B to minimize the number of days to recovery:

```
5     REM MINIMIZE D THE NUMBER
6     REM OF DAYS TO RECOVERY
7     REM FOR CONDITION K PATIENTS
10    B=999999
20    FOR X1=0 TO 7
30    FOR X2=0 TO 7
40    D1=20+X1**2-8*X1
50    D2=X2**2-12*X2-.10*X1*X2
60    D=33+D1+D2
70    IF D < B THEN 90
80    GO TO 120
90    A1=X1
100   A2=X2
110   B=D
120   NEXT X2
130   NEXT X1
140   PRINT A1,A2
150   PRINT D
160   STOP
170   END

4              6
6.10000
```

Therefore, 4 units of Drug A and 6 units of Drug B will lead to a recovery in 6.1 days on the average. This is the best solution (least number of days).

EXAMPLE 6.2

Patients with a particular condition have to take medicine each day for the rest of their lives. Therefore, it was decided to try to improve the taste of it. Twenty different compounds consisting of the medicine and various amounts of two additives (x_1 and x_2) and a sweetener (x_3) were given to patients. They were asked to rate the taste as 20% better or 10% worse, for example, than their standard compound.

From these taste test results a predictive taste curve as a function of x_1, x_2, and x_3 was developed statistically: $P = -x_1{}^2 + 60x_1 - 1.1x_2{}^2 + 1000 - .009x_1x_3 + 40x_2 - .86x_3{}^2 + 90x_3 + .006x_1x_2 + .01x_2x_3$ subject to $0 \leqslant x_i \leqslant 95$ units for $i = 1, 2, 3$.

Therefore, we write a Monte Carlo simulation program to try to find the optimal solution:

```
5      REM IMPROVING TASTE OF MEDICINE
10     REM WITH RIGHT COMBINATION
15     REM OF TWO ADDITIVES AND A
20     REM SWEETENER.  THIS IS NEEDED
30     REM BECAUSE PATIENTS MUST TAKE
40     REM THIS MEDICINE FOR THE
50     REM REST OF THEIR LIVES.
60     X=1
70     B=-999999
80     FOR I=1 TO 10000
90     X1=INT(RND(X)*96)
100    X2=INT(RND(X)*96)
110    X3=INT(RND(X)*96)
120    PO=1000-.009*X1*X3
130    P1=-X1**2+60*X1
140    P2=-1.10*X2**2+40*X2
150    P3=-.86*X3**2+90*X3
160    P4=.006*X1*X2+.01*X2*X3
170    P=PO+P1+P2+P3+P4
180    IF P > B THEN 200
190    GO TO 240
200    A1=X1
210    A2=X2
220    A3=X3
230    B=P
240    NEXT I
250    PRINT A1,A2,A3
260    PRINT B
270    STOP
280    END
```

```
28              19              52
4613.43

27              17              52
4606.62
```

```
29              20              51
4612.51

29              20              52
4613.87

28              19              50
4608.99
```

From the printouts it looks like x_1 is between 24 and 34, x_2 is between 14 and 24, and x_3 is between 46 and 56. So we modify the program to search these 11 x 11x 11 = 1331 ordered triples for the optimal solution. The program is:

```
5     REM IMPROVING TASTE OF MEDICINE
10    REM WITH RIGHT COMBINATION
15    REM OF TWO ADDITIVES AND A
20    REM SWEETENER.  THIS IS NEEDED
30    REM BECAUSE PATIENTS MUST TAKE
40    REM THIS MEDICINE FOR THE
50    REM REST OF THEIR LIVES.
60    X=1
70    B=-999999
80    FOR X1=24 TO 34
90    FOR X2=14 TO 24
100   FOR X3=46 TO 56
120   P0=1000-.009*X1*X3
130   P1=-X1**2+60*X1
140   P2=-1.10*X2**2+40*X2
150   P3=-.86*X3**2+90*X3
160   P4=.006*X1*X2+.01*X2*X3
170   P=P0+P1+P2+P3+P4
180   IF P > B THEN 200
190   GO TO 240
200   A1=X1
210   A2=X2
220   A3=X3
230   B=P
240   NEXT X3
241   NEXT X2
242   NEXT X1
250   PRINT A1,A2,A3
260   PRINT B
270   STOP
280   END

30          18          52
4616.72
```

Therefore, 30 units of Additive 1, 18 units of Additive 2, and 52 units of the sweetener should improve the taste by about 46%. So this compound (and maybe a few like it) should be made up and taste tested.

EXAMPLE 6.3

Two drugs have been somewhat effective separately in reducing high blood pressure. So an experiment was performed to see if better results could be obtained by treating patients with a combination of the drugs. The dosage of Drug 1 (x_1) was restricted to 20 units. The dosage of Drug 2 (x_2) was restricted to 15 units. Also, the number of times the drugs were given per week (x_3) was varied and restricted to no more than 28 times. After tests with several patients, the response curve was determined to be $P = -x_1{}^2 + 22x_1 - 6.25x_2{}^2 + 100x_2 - x_3{}^2 + 30x_3 + 1000 + .002x_1x_2x_3 + .03x_2x_3 + .06x_1x_3$.

We write a program to maximize the function (in this case, this will maximize the reduction in blood pressure):

```
5     REM MAXIMIZE PERCENTAGE REDUCTION
10    REM OF BLOOD PRESSURE AS A FUNCTION
15    REM OF DOSAGE OF TWO DRUGS AND
20    REM FREQUENCY OF TAKING THE MEDICINE
25    B=-999999
30    FOR X1=0 TO 20
40    FOR X2=0 TO 15
50    FOR X3=0 TO 28
60    P0=1000+.002*X1*X2*X3
70    P1=-X1**2+22*X1
80    P2=-6.25*X2**2+100*X2
90    P3=-X3**2+30*X3
100   P4=.03*X2*X3+.06*X1*X3
110   P=P0+P1+P2+P3+P4
120   IF P > B THEN 140
130   GO TO 175
140   A1=X1
150   A2=X2
160   A3=X3
170   B=P
175   NEXT X3
176   NEXT X2
177   NEXT X1
180   PRINT A1,A2,A3
190   PRINT B
200   STOP
210   END
```

```
12              8              16
1762.43
```

Therefore, 12 units of Drug 1 and 8 units of Drug 2 given 16 times per week yielded a maximal reduction in blood pressure of 17.6243% (the yields were in hundredths of a percent).

However, one of the physicians involved in the experiment is concerned that 12 units plus 8 units might be too much for some patients' systems and proposed that a search for the optimal solution when $x_1 + x_2$ is less than or equal to 16 units be tried. The program is below (note the constraint is in line 55):

```
5    REM MAXIMIZE PERCENTAGE REDUCTION
10   REM OF BLOOD PRESSURE AS A FUNCTION
15   REM OF DOSAGE OF TWO DRUGS AND
20   REM FREQUENCY OF TAKING THE MEDICINE
25   B=-999999
30   FOR X1=0 TO 20
40   FOR X2=0 TO 15
50   FOR X3=0 TO 28
55   IF X1+X2 > 16 THEN 175
60   P0=1000+.002*X1*X2*X3
70   P1=-X1**2+22*X1
80   P2=-6.25*X2**2+100*X2
90   P3=-X3**2+30*X3
100  P4=.03*X2*X3+.06*X1*X3
110  P=P0+P1+P2+P3+P4
120  IF P > B THEN 140
130  GO TO 175
140  A1=X1
150  A2=X2
160  A3=X3
170  B=P
175  NEXT X3
176  NEXT X2
177  NEXT X1
180  PRINT A1,A2,A3
190  PRINT B
200  STOP
210  END

8          8            15
1749.72
```

Therefore, 8 units of each drug given 15 times a week satisfies the constraint and would result in about a 17.4972% reduction in blood pressure. This is just about as good as the previous solution.

EXAMPLE 6.4

One medical research team has been experimenting with a combination of exercise and a drug to reduce blood pressure in hypertensive patients. The first variable, x_1, is the number of tenths of kilometers run per day by their patients (up to a maximum of 150, or 15 kilometers). The second variable, x_2, is the number of units of Drug L taken per day, up to a maximum of 100. The response curve from the data on the patients is $P = -x_1^2 + 76x_1 - 2.25x_2^2 + 210x_2 - 6323$, where P is the drop in blood pressure in points. We write a program to maximize this drop:

```
5     REM MAXIMIZE REDUCTION IN
10    REM BLOOD PRESSURE (IN POINTS)
20    REM AS FUNCTION OF NUMBER
30    REM OF TENTHS OF KILOMETERS
40    REM RUN PER DAY (X1) AND UNITS
50    REM OF DRUG L PER DAY TAKEN.
60    B=-999999
70    FOR X1=0 TO 150
80    FOR X2=0 TO 100
90    P1=1000-X1**2+76*X1
100   P2=-2.25*X2**2+210*X2
110   P=P1+P2-7323
120   IF P > B THEN 140
130   GO TO 170
140   A1=X1
150   A2=X2
160   B=P
170   NEXT X2
180   NEXT X1
190   PRINT A1,A2
200   PRINT B
210   STOP
220   END

38              47
20.7500
```

Therefore, 3.8 kilometers a day and 47 units of Drug L per day should drop the high blood pressure by 20.75 points on the average. More testing should be done near this optimal solution, and perhaps other variables introduced.

EXAMPLE 6.5

Doctors have been trying to reduce the size of malignant tumors through treatment with combinations of Drug K (x_1 units), Drug M (x_2 units), and various levels of cobalt (x_3 units).

After many tests the response curve was modeled statistically as $P = 3x_1{}^2 + 7x_2{}^2 + 2.1x_3{}^2 + 6x_1 + 5x_2 + 14x_3 - x_1x_2 + 2x_1x_3 - 1.6x_2x_3 + .49x_1x_2x_3$ subject to $0 \leqslant x_1 \leqslant 75$, $0 \leqslant x_2 \leqslant 70$, $0 \leqslant x_3 \leqslant 20$, and $x_1 + x_2 \leqslant 100$ and $x_2 + x_3 \leqslant 50$. These last two constraints are designed to reduce and prevent harmful and painful side effects from overdosage. The program to maximize the function is:

```
5     REM MAXIMIZING REDUCTION IN
10    REM TUMOR SIZE WITH DRUG K,
15    REM DRUG M AND LEVEL
20    REM OF COBALT TREATMENT.
30    B=-999999
40    FOR X1=0 TO 75
50    FOR X2=0 TO 70
60    FOR X3=0 TO 20
70    P1=3*X1**2+7*X2**2+2.1*X3**2
80    P2=6*X1+5*X2+14*X3
90    P3=-X1*X2+2*X1*X3-1.6*X2*X3
100   P4=.49*X1*X2*X3
110   P=P1+P2+P3+P4
65    IF X1+X2 > 100 THEN 180
66    IF X2+X3 > 50 THEN 180
120   IF P > B THEN 140
130   GO TO 180
140   A1=X1
150   A2=X2
160   A3=X3
170   B=P
180   NEXT X3
190   NEXT X2
200   NEXT X1
210   PRINT A1,A2,A3
220   PRINT B
230   STOP
240   END

70          30          20
43010.0
```

Therefore, 70 units of Drug K, 30 units of Drug M, and 20 doses of cobalt treatment will lead to a maximum reduction of 43.01% (the response curve variable *P* gives answers in thousandths of a percent).

EXAMPLE 6.6

The Sports Medicine Clinic has been experimenting with developing distance runners as a function of daily running distance x_1, sleep x_2, weight x_3, and age x_4. With various combinations of the four, the athlete's pulse is taken at the end of a year of sticking to their plan of x_1, x_2, x_3, and x_4. The idea is that the lower the pulse, the better condition the runner is in. The actual response term is pulse reduction in units.

The response curve modeled statistically from the data was $P = 1000 - .001x_1x_2x_3x_4 - x_1^2 + 50x_1 - 2x_2^2 + 1500x_2 - x_3^2 - 40x_3 - 3x_3^2 + 70x_4 + .01x_1x_2 - .1x_1x_3 + .2x_1x_4 - .02x_2x_3 + .15x_2x_4 - .03x_3x_4 - .06x_1x_2x_3 + .0001x_1x_2x_4 - .055x_1x_3x_4 - .002x_2x_3x_4$ subject to $0 \leqslant x_1 \leqslant 40$ kilometers per day of running, $360 \leqslant x_2 \leqslant 540$ minutes of sleep per night, $-30 \leqslant x_3 \leqslant 10$ percentage of weight above or below the normal for the individuals' height and age.

If an athlete aged 19 comes in for the optimal running training prescription, what should it be? The program follows:

```
5    REM DEVELOPING A DISTANCE RUNNER
10   REM BY MODELING CONDITIONING AS A
15   REM FUNCTION OF DAILY RUNNING DISTANCE.
20   REM SLEEP, WEIGHT AND AGE, THIS
25   REM PARTICULAR OPTIMAL TRAINING
30   REM PLAN IS FOR AGE (X4) FIXED AT 19.
40   B=-999999
50   FOR X1=0 TO 40
60   FOR X2=360 TO 540
70   FOR X3=-30 TO 10
80   X4=19
90   P0=1000-.001*X1*X2*X3*X4
100  P1=-X1**2+50*X1
110  P2=-2*X2**2+1500*X2
120  P3=-X3**2-40*X3
130  P4=-3*X3**2+70*X4
140  P5=.01*X1*X2-.1*X1*X3
150  P6=.2*X1*X4-.02*X2*X3
160  P7=.15*X2*X4-.03*X3*X4
170  P8=-.06*X1*X2*X3+.0001*X1*X2*X4
180  P9=-.005*X1*X3*X4-.002*X2*X3*X4
190  P=P1+P2+P3+P4+P5+P6+P7+P8+P9+P0
200  IF P > B THEN 220
210  GO TO 270
220  A1=X1
230  A2=X2
```

```
240  A3=X3
250  A4=X4
260  B=P
270  NEXT X3
280  NEXT X2
290  NEXT X1
300  PRINT A1,A2,A3
310  PRINT B
320  STOP
330  END

40          400          -30
320679.
```

Therefore at age 19, 40 kilometers per day, 400 minutes of sleep per night, and 30% below the average weight for the individuals' height should be goals which should reduce the runner's pulse about 32,0679 units (pulse reduction was in ten thousandths of a unit).

EXAMPLE 6.7

Patients have a certain condition that will cure itself in time. The physicians goal, then, is to reduce the pain of the symptoms by treating the patients with a combination of Drug 1 and Drug 2 in total units not to exceed 95. The response curve is $P = -6620 - 2x_1{}^2 + .8x_2{}^2 + 40x_1 + .5x_2 + .09x_1x_2$ subject to $0 \leqslant x_1 \leqslant 40$ and $0 \leqslant x_2 \leqslant 90$, where P is the percentage of improvement the patients say they feel (on a scale of 0 to 100 percent) after the drug treatment.

```
5    REM MAXIMIZE REDUCTION IN PAIN
10   REM BY TREATING SYMPTOMS
15   REM WITH A COMBINATION OF DRUG 1 AND
20   REM DRUG 2 (NOT TO EXCEED 95 UNITS)
30   B=-999999
40   FOR X1=0 TO 40
50   FOR X2=0 TO 90
60   IF X1 + X2 > 95 THEN 150
70   P1=80-2*X1**2+.8*X2**2
80   P2=40*X1+.5*X2+.09*X1*X2
90   P=P1+P2-6700
100  IF P > B THEN 120
110  GO TO 150
120  A1=X1
130  A2=X2
140  B=P
```

```
150  NEXT X2
160  NEXT X1
170  PRINT A1,A2
180  PRINT B
190  STOP
200  END

5             90
95.5000
```

Therefore, $x_1 = 5$ units of Drug 1 and $x_2 = 90$ units of Drug 2 should lead to about a 95% improvement in how the patients feel.

EXAMPLE 6.8

Researchers have been studying the population of a large country to see if they can develop a profile of the environmental circumstances that would tend to reduce the chance of the average person getting a particular disease. Preliminary statistical analysis yields four promising independent variables:

- x_1 Town size in units of 10,000 with $x_1 = 100$ representing any town with over one million people
- x_2 Level of exercise on a scale of 0 to 100
- x_3 Diet containing 10 to 40 units of a particular vitamin in the daily diet
- x_4 Level of drinking 0 to 100 (0 is none, 100 is moderate)

The response curve was percentage chance of getting the disease (where 100% is considered average).

The equation was $P = 925 - .5x_1x_4 + x_2^2 - 10x_2 + x_3^2 - 56x_3 + x_4^2 - 20x_4$ subject to $0 \leqslant x_1 \leqslant 100$, $0 \leqslant x_2 \leqslant 100$, $10 \leqslant x_3 \leqslant 40$, and $0 \leqslant x_4 \leqslant 100$. The program follows:

```
5     REM MINIMIZE THE CHANCE OF
10    REM DEVELOPING DISEASE R BY
15    REM LIVING IN A SMALL TOWN
20    REM PROPER EXERCISE AND DIET
25    REM AND LIGHT TO MODERATE DRINKING
27    X=1
30    B=-999999
40    FOR I=1 TO 10000
50    X1=INT(RND(X)*101)
60    X2=INT(RND(X)*101)
```

```
70   X3=10+INT(RND(X)*31)
80   X4=INT (RND(X)*101)
90   P0=925-.5*X1*X4
100  P2=X2**2-10*X2
120  P3=X3**2-56*X3
130  P4=X4**2-20*X4
140  P=P0+P1+P2+P3+P4
150  IF P > B THEN 170
160  GO TO 220
170  A1=X1
180  A2=X2
190  A3=X3
200  A4=X4
210  B=P
220  NEXT I
230  PRINT A1,A2,A3,A4
240  PRINT B
250  STOP
260  END
```

```
18          100         20          97
15101

0            98         34         100
16101

1           100         35          95
15567.5

0           100         14          99
16458

0            98         13         100
16290

0           100         25         100
16450

0           100         10         100
16765

0           100         10         100
16765

0           100         10         100
16765
```

From these Monte Carlo runs it was decided to fix $x_2 = 100$ and $x_4 = 100$ and rerun. The result was:

$x_1 = 0$ (live in a small town)
$x_2 = 100$ (a lot of exercise)
$x_3 = 10$ (units of the vitamin)
$x_4 = 100$ (moderate drinking)

can reduce one's chances of contracting the disease to 16.765% (response values in thousandths of a percent) of the national average.

EXERCISES

6.1 Maximize $P = 1000 - x_1{}^2 + 37x_1 - 2x_2{}^2 + 101x_2 + .2x_1x_2$ subject to $0 \leqslant x_1 \leqslant 100$, $0 \leqslant x_2 \leqslant 100$, and $x_1 + x_2 \leqslant 150$, where x_1 is the number of units of Drug A, x_2 is the number of units of Drug B, and $x_1 + x_2$ must be less than 150 units per day, or dangerous side effects could begin. P is the perceived pain reduction by a patient who has a condition that will run its course. Therefore, the object is to make the patient as comfortable as possible during the period of illness.

6.2 Maximize $P = 2009 + 2x_1 + 5x_2 + 6x_3 + 2x_1x_2 + 3x_1x_3 + 2.2x_2x_3 + .25 x_1x_2x_3$ subject to $0 \leqslant x_1 \leqslant 100$, $0 \leqslant x_2 \leqslant 100$, and $0 \leqslant x_3 \leqslant 100$. The x_i's are daily dosages of three drugs, P is the percentage reduction in tumor size (in hundredths of a percent), and $x_1 + 2x_2 + 3x_3 \leqslant 200$ and $2x_1 + 1.5x_2 + x_3 \leqslant 150$, or the patient will have terrible side effects and suffer undue pain.

6.3 Maximize $P = 10{,}000 - 2x_1{}^2 + 47x_1 - 8x_2{}^2 + 306x_2 - 6x_3{}^2 + 100x_3 - 2x_4{}^2 + 205xy - 8x_5{}^2 + 104x_5 + x_1x_2 + x_2x_3 + 3x_4x_5 - x_1x_2x_3 - x_4x_5 + .009 x_1x_2x_3x_5$ subject to $0 \leqslant x_1 \leqslant 100$, $0 \leqslant x_1 \leqslant 14$, $0 \leqslant x_3 \leqslant 100$, $0 \leqslant x_4 \leqslant 14$, and $0 \leqslant x_5 \leqslant 12$. x_1 is the number of units of Drug 1 given to the patient; x_2 is the number of times the dosage of Drug 1 is given to the patient per week; x_3 is the number of units of Drug 2 given to the patient; x_4 is the number of times the dosage of Drug 2 is given to the patient per week; x_5 is how many kilometers the patient runs per day. P is the percentage reduction in pulse rate (in thousandths of a percent) for hypertensive patients.
Hint: Do a simulation and the focus for the optimum.

6.4 Maximize $P = 2050 - x_1{}^2 + 75x_1 - 2.6x_2{}^2 + 109x_2 + 2.1x_1x_2 - x_3{}^2 + 111x_3 + .008x_1x_3 + .02x_2x_3$ subject to $0 \leqslant x_1 \leqslant 1000$, $0 \leqslant x_2 \leqslant 1000$, and $0 \leqslant x_3 \leqslant 12$. x_1 is the daily dosage of Drug 1, x_2 is the daily dosage of Drug 2, and x_3 is the number of hours between the administering of drug 1 and drug 2. P is the percentage reduction in blood pressure in hundredths of a percent.

Suggested Reading

Beltrami, Edward. *Models for Public Systems Analysis.* New York: Academic Press, 1977.

Conley, William. *Making a Prescription Drug.* International Journal of Mathematical Education in Science and Technology, 1981.

Cress; Dirksen; and Graham. *Fortran IV with WATFOR & WATFIV.* Englewood Cliffs, N.J.: Prentice-Hall, 1973.

Goldstein; Loy; and Schneider. *Calculus and Its Applications.* Englewood Cliffs, N.J.: Prentice-Hall, 1977.

Sass, Joseph. *Fortran IV Programming and Applications.* San Francisco: Holden-Day, 1974.

Timbergen, J. "Effects of the Computerization of Research, COMPSTAT 1978 Proceedings," in *Computational Statistics*, pp. 20-28. Wurzburg, Vienna: Physica-Verlag, 1978.

CHAPTER 7

Financial Planning

What are the key factors in budgeting for the manufacture of a product? There are many answers to this question, but one of the most important factors is to estimate demand as a function of price (and other factors) and balance this with the cost of capital and discounts for buying supplies in quantity. This chapter is devoted to these types of financial planning problems. Appendix A presents a few ideas on estimating demand curves.

EXAMPLE 7.1

The Silver Bay Company is thinking of producing two products in its north building. Marketing research has established a price-quantity curve for Product 1 of $P_1 = -.2x_1 + 950$, where x_1 is the number of units sold at price P_1. The unit cost for Product 1 is \$50 and the tax savings (due to a government incentive plan) is \$1000. Marketing has estimated a price-quantity curve for Product 2 to be $P_2 = -.15x_2 + 1200$, where x_2 is the number of units sold at price P_2. The unit cost for this product is \$55 and the tax savings for helping to supply this needed product is \$1500. Product 1 takes 10 hours of labor per unit, while Product 2 takes 12 hours of labor per unit. There are 40,000 worker-hours available. Therefore, we would like to maximize $P = (-.2x_1 + 950)x_1 - 50x_1 + 1000y_1 + (-.15x_2 + 1200)x_2 - 55x_2 + 1500y_2$ subject to $x_1 \geqslant 0$, $x_2 \geqslant 0$, $10x_1 + 12x_2 \leqslant 40{,}000$, and $y_1 = 1$ if $x_1 > 0$ and $y_2 = 1$ if $x_2 > 0$, and $y_1 = 0$ if $x_1 = 0$ and $y_2 = 0$ if $x_2 = 0$.

The program follows, searching for the maximum profit solution by increments of 10 (note that the tax breaks are dealt with in lines 80, 90, 100, 110, 115 and 120):

```
5     REM PROFIT MAXIMIZATION WITH
10    REM FIXED COSTS, DEMAND CURVES,
20    REM UNIT COSTS, AND LABOR CONSTRAINTS
30    B=-999999
40    FOR X1=0 TO 4000 STEP 10
50    FOR X2=0 TO 3333 STEP 10
60    IF 10*X1+12*X2 > 40000 THEN 190
70    P=0
80    IF X1 > 0 THEN 100
90    GO TO 110
100   P=P+1000
110   IF X2 > 0 THEN 120
115   GO TO 125
120   P=P+1500
125   P1=-.2*X1**2+950*X1
130   P2=-.15*X2**2+1200*X2
135   P=P+P1+P2
136   P=P-50*X1-55*X2
140   IF P > B THEN 160
150   GO TO 190
160   A1=X1
170   A2=X2
180   B=P
190   NEXT X2
200   NEXT X1
210   PRINT A1, A2
220   PRINT B
230   STOP
240   END

1300          2250
2.65138E+06
```

Therefore, 1300 units of Product 1 and 2250 units of Product 2 would yield $2,651,380 profit.

Let us focus our search for the optimum to between 1200 and 1400 for x_1 and between 2150 and 2350 for x_2. The rest of the program is the same as before:

```
4      REM FOCUS SEARCH AFTER FIRST RUN
5      REM PROFIT MAXIMIZATION WITH
10     REM FIXED COSTS, DEMAND CURVES,
20     REM UNIT COSTS, AND LABOR CONSTRAINTS
30     B=-999999
40     FOR X1=1200 TO 1400
50     FOR X2=2150 TO 2350
60     IF 10*X1+12*X2 > 40000 THEN 190
70     P=0
80     IF X1 > 0 THEN 100
90     GO TO 110
100    P=P+1000
110    IF X2 > 0 THEN 120
115    GO TO 125
120    P=P+1500
125    P1=-.2*X1**2+950*X1
130    P2=-.15*X2**2+1200*X2
135    P=P+P1+P2
136    P=P-50*X1-55*X2
140    IF P > B THEN 160
150    GO TO 190
160    A1=X1
170    A2=X2
180    B=P
190    NEXT X2
200    NEXT X1
210    PRINT A1, A2
220    PRINT B
230    STOP
240    END

1282           2265
2.65149E+06
```

Therefore, $x_1 = 1282$ units of Product 1 and $x_2 = 2265$ units of Product 2 yields the maximum profit of $2,651,490.

However, the plant manager of the north building feels that the output figures are underestimated and claims that as the workers make more and more units of products 1 and 2 their efficiency will improve. Specifically, it will not take 10 hours per unit on Product 1 but $10x_1^{.95}$ (less average time to make the product as the workers make more). And on Product 2 it will be $12x_2^{.88}$ hours for production level x_2.

Therefore, $10x_1 + 12x_2 \leqslant 40{,}000$ should be replaced by $10x_1^{.95} + 12x_2^{.88} \leqslant 40{,}000$. The program below reflects that change. Also note that in lines 40 and 50, x_1 and x_2, respectively, have new bounds. This time we used the fact that $P_1 = 0$ when $(-.2x_1 + 950)x_1 = 0$. Therefore, when $x_1 = 4500$, $P_1 = 0$. So $x_1 \leqslant 4500$. Similarly, $P_2 = 0$ when $(-.15x_2 + 1200)x_2 = 0$. Therefore, when $x_2 = 8000$, $P_2 = 0$. So $x_2 \leqslant 8000$. Actually, lower bounds could have been found.

```
5     REM PROFIT MAXIMIZATION WITH
10    REM FIXED COSTS, DEMAND CURVES,
20    REM UNIT COSTS, AND LABOR CONSTRAINTS
25    REM WITH EFFICIENCIES FOR SCALE
30    B=-999999
40    FOR X1=0 TO 4500 STEP 20
50    FOR X2=0 TO 8000 STEP 20
60    IF 10*X1**.95+12*X2**.88 > 40000 THEN 190
70    P=0
80    IF X1 > 0 THEN 100
90    GO TO 110
100   P=P+1000
110   IF X2 > 0 THEN 120
115   GO TO 125
120   P=P+1500
125   P1=-.2*X1**2+950*X1
130   P2=-.15*X2**2+1200*X2
135   P=P+P1+P2
136   P=P-50*X1-55*X2
140   IF P > B THEN 160
150   GO TO 190
160   A1=X1
170   A2=X2
180   B=P
190   NEXT X2
200   NEXT X1
210   PRINT A1, A2
220   PRINT B
230   STOP
240   END

2240              3820
3.20002E+06
```

2,240 units of Product 1 and 3820 units of Product 2 will yield \$3,200,020 in profit.

The focus search below centers on $2140 \leqslant x_1 \leqslant 2340$ and $3720 \leqslant x_2 \leqslant 3920$:

```
4     REM FOCUS SEARCH AFTER FIRST RUN
5     REM PROFIT MAXIMIZATION WITH
10    REM FIXED COSTS, DEMAND CURVES,
20    REM UNIT COSTS, AND LABOR CONSTRAINTS
25    REM WITH EFFICIENCIES FOR SCALE
30    B=-999999
40    FOR X1=2140 TO 2340
50    FOR X2=3720 to 3920
60    IF 10*X1**.95+12*X2**.88 > 40000 THEN 190
70    P=0
80    IF X1 > 0 THEN 100
90    GO TO 110
100   P=P+1000
110   IF X2 > 0 THEN 120
115   GO TO 125
120   P=P+1500
125   P1=-.2*X1**2+950*X1
130   P2=-.15*X2**2+1200*X2
135   P=P+P1+P2
136   P=P-50*X1-55*X2
140   IF P > B THEN 160
150   GO TO 190
160   A1=X1
170   A2=X2
180   B=P
190   NEXT X2
200   NEXT X1
210   PRINT A1, A2
220   PRINT B
230   STOP
240   END

2250              3817
3.20004E+06
```

Therefore, $x_1 = 2250$ and $x_2 = 3817$ yields maximum profit of \$3,200,040. Also, substituting $x_1 = 2250$ in $P_1 = -.2x_1 + 950$ yields P_1 = \$500, which is the price Silver Bay should charge for one unit of Product 1. Similarly, substituting $x_2 = 3817$ into $P_2 = -.15x_2 + 1200$ yields P_2 = \$627.45, which is what they should charge per unit for Product 2.

Also note that the 2250 units plus 3817 units only uses $10 \times 2250^{.95} + 12 \times 3817^{.88} = 32{,}321$ hours of worker time out of the 40,000 hours available. Therefore, reassignment of workers could also cut costs, etc.

EXAMPLE 7.2

The Mackinac Island Company buys unfinished brass sections, finishes them into polished brass fittings, and then resells them. The price-quantity curve has been determined to be $P_1 = -.0005x_1 + 25$, where x_1 is the number of units sold per month. A maximum of 60,000 per month can be sold. The problem is that the company would like to buy the brass sections in quantity because of the substantial discounts for doing so. But they must balance this with the fact that they will have to borrow the money at 11% in order to buy in quantity. The price list for the brass section is as follows:

Number of units	Unit price ($)
0 - 4,999	5.00
5,000 - 24,999	4.80
25,000 - 49,999	4.60
50,000 - 99,999	4.50
100,000 - 199,999	4.25
200,000 - 299,999	4.00
300,000 - 499,999	3.50
500,000 or more	3.00

Therefore, we must write a program to maximize $P = ((-.0005x_1 + 25)x_1 - CX_1)12 - 12x_1\ C(.11)$ subject to $0 \leqslant x_1 \leqslant 60{,}000$, where C is the unit price at the quantity level arrived at. Notice in the program below that lines 95 through 250 deal with getting the right C value (unit cost) for each x_1, as x_1 goes from 0 to 60,000:

```
5     REM PROFIT MAXIMIZATION WITH
10    REM DISCOUNTS FOR BUYING IN
15    REM QUANTITY, BUT HAVING TO
20    REM BORROW THE MONEY TO BUY,
30    REM ALSO CONSIDERING DEMAND AS
35    REM A FUNCTION OF PRICE.
85    B=-999999
90    FOR X1=0 TO 60000
95    IF 12*X1 < 5000 THEN 190
100   IF 12*X1 < 25000 THEN 200
110   IF 12*X1 < 50000 THEN 210
120   IF 12*X1 < 100000 THEN 220
130   IF 12*X1 < 200000 THEN 230
140   IF 12*X1 < 300000 THEN 240
150   IF 12*X1 < 500000 THEN 250
```

```
160  C=3.00
165  GO TO 260
190  C=5.00
195  GO TO 260
200  C=4.80
205  GO TO 260
210  C=4.60
215  GO TO 260
220  C=4.50
225  GO TO 260
230  C=4.25
235  GO TO 260
240  C=4.00
245  GO TO 260
250  C=3.50
260  P1=(-.0005*X1**2+25*X1-C*X1)*12
265  P2=(12*X1*C)*.11
270  P=P1-P2
290  IF P > B THEN 310
300  GO TO 330
310  A1=12*X1
315  C1=C
316  K1=P1
317  K2=P2
320  B=P
330  NEXT X1
340  PRINT A1,B
345  PRINT C1
346  PRINT K1,K2
350  STOP
360  END

300000              2584500
3.50000
2700000             115500
```

Therefore, the company should purchase, finish, and sell 300,000 brass sections. They will pay \$3.50 per section, or \$1,050,000 for this inventory. Therefore, the revenue, \$2,700,000, minus the interest will yield \$2,584,500 profit (if the demand curve is approximately right).

Also, substituting 25,000 = (300,000/12) into the monthly demand curve $P_1 = -.0005x_1 + 25x_1$ yields $P_1 = \$12.50$. Therefore, the company should charge \$12.50 per unit on their finished product.

However, just before closing the deal, the company learns that they can borrow the money, but at 12%, not 11%. Therefore, they rerun the program (only line 265 changes) as follows.

```
5     REM PROFIT MAXIMIZATION WITH
10    REM DISCOUNTS FOR BUYING IN
15    REM QUANTITY, BUT HAVING TO
20    REM BORROW THE MONEY TO BUY.
30    REM ALSO CONSIDERING DEMAND AS
35    REM A FUNCTION OF PRICE.
40    REM NEW SOLUTION OF INTEREST
45    REM RATE RISES FROM .11 to .12
85    B=-999999
90    FOR X1=0 TO 60000
95    IF 12*X1 < 5000 THEN 190
100   IF 12*X1 < 25000 THEN 200
110   IF 12*X1 < 50000 THEN 210
120   IF 12*X1 < 100000 THEN 220
130   IF 12*X1 < 200000 THEN 230
140   IF 12*X1 < 300000 THEN 240
150   IF 12*X1 < 500000 THEN 250
160   C=3.00
165   GO TO 260
190   C=5.00
195   GO TO 260
200   C=4.80
205   GO TO 260
210   C=4.60
215   GO TO 260
220   C=4.50
225   GO TO 260
230   C=4.25
235   GO TO 260
240   C=4.00
245   GO TO 260
250   C=3.50
260   P1=(-.0005*X1**2+25*X1-C*X1)*12
265   P2=(12*X1*C)*.12
270   P=P1-P2
290   IF P > B THEN 310
300   GO TO 330
310   A1=12*X1
315   C1=C
316   K1=P1
317   K2=P2
320   B=P
330   NEXT X1
340   PRINT A1,B
345   PRINT C1
346   PRINT K1,K2
350   STOP
360   END
300000           2.57400E+06
3.50000
2700000          126000.
```

The solution is still the same—300,000 units purchased at $3.50 finished and resold at $12.50. The only change is that the increased interest drops the profit to $2,574,000.

Further, the Mackinac Island Company is considering a deal on chrome fittings from a different supplier. Their estimated price-quantity curve is $P_2 = -.0009x_2 + 28.80$, where x_2 is the number of units sold per month and P_2 is the price. The price list for chrome is as follows:

Number of units	Unit price ($)
0 - 9,999	6.00
10,000 - 29,999	5.60
30,000 - 74,999	5.20
75,000 - 199,999	5.00
200,000 - 349,999	4.45
350,000 - 499,999	4.25
500,000 or more	4.00

Again, they will borrow the money at 12% to buy this inventory. They believe that demand will not exceed 38,888 per month. So the problem is to maximize $P = ((-.0009x_2 + 28.80)x_2 - C_2X_2)12 - 12x_2C_2(.12)$, where x_2 is the number of units sold per month and C_2 is the unit price level (depending on the amount purchased). The .12 is for the 12% rate on the loan, and the other 12's are for the twelve months in the year as in the previous problem. The program is below:

```
5    REM PROFIT MAXIMIZATION WITH
10   REM DISCOUNTS FOR BUYING IN
15   REM QUANTITY, BUT HAVING TO
20   REM BORROW THE MONEY TO BUY,
30   REM ALSO CONSIDERING DEMAND
35   REM AS A FUNCTION OF PRICE.
85   B=-999999
90   FOR X2=0 TO 38888
95   IF 12*X2 < 10000 THEN 190
100  IF 12*X2 < 30000 THEN 200
110  IF 12*X2 < 75000 THEN 210
120  IF 12*X2 < 200000 THEN 220
130  IF 12*X2 < 350000 THEN 230
140  IF 12*X2 < 500000 THEN 240
160  C2=4.00
165  GO TO 260
190  C2=6.00
195  GO TO 260
200  C2=5.60
205  GO TO 260
```

```
210  C2=5.20
215  GO TO 260
220  C2=5.00
225  GO TO 260
230  C2=4.45
235  GO TO 260
240  C2=4.25
260  P3=(-.0009*X2**2+28.80*X2-C2*X2)*12
270  P4=(12*X2*C2)*.12
280  P5=P3-P4
290  IF P5 > B THEN 310
300  GO TO 330
310  A2=12*X2
315  C5=C2
316  K3=P3
317  K4=P4
320  B=P5
330  NEXT X2
340  PRINT A2,B
350  PRINT C5
355  PRINT K3,K4
360  STOP
370  END

154668               1.79413E+06
5
1.88693E+06          92800.8
```

Therefore, they should borrow \$773,340 to buy 154,668 units of chrome to maximize profit—\$1,794,130 in this case.

In order to purchase both inventories the company must borrow \$773,340 + 1,050,000 = \$1,823,340. However, they receive some bad news from the bank. The bank will only loan them \$1,250,000 at 12%. Therefore, the company will combine the two previous problems and run one program with the constraint $12cx_1 + 12c_2x_2 \leqslant \$1,250,000$, where c and c_2 are the price levels of the two products. The program is below:

```
1    REM TWO PRODUCT PROBLEM
2    REM YOU NEED TO BORROW
3    REM 1,800,000 BUT BANK
4    REM WILL LOAN ONLY 1,250,000
5    REM PROFIT MAXIMIZATION WITH
10   REM DISCOUNTS FOR BUYING IN
```

```
15    REM QUANTITY, BUT HAVING TO
20    REM BORROW THE MONEY TO BUY.
30    REM ALSO CONSIDERING DEMAND
35    REM AS A FUNCTION OF PRICE.
36    REM FOCUS SEARCH
37    B=-999999
40    FOR X1=0 TO 60000 STEP 100
45    FOR X2=0 TO 38888 STEP 100
47    IF 12*X1 < 5000 THEN 65
49    IF 12*X1 < 25000 THEN 67
51    IF 12*X1 < 50000 THEN 69
53    IF 12*X1 < 100000 THEN 71
55    IF 12*X1 < 200000 THEN 73
57    IF 12*X1 < 300000 THEN 75
59    IF 12*X1 < 500000 THEN 77
61    C=3.00
62    GO TO 95
65    C=5.00
66    GO TO 95
67    C=4.80
68    GO TO 95
69    C=4.60
70    GO TO 95
71    C=4.50
72    GO TO 95
73    C=4.25
74    GO TO 95
75    C=4.00
76    GO TO 95
77    C=3.50
95    IF 12*X2 <  10000 THEN 190
100   IF 12*X2 <  30000 THEN 200
110   IF 12*X2 <  75000 THEN 210
120   IF 12*X2 <  200000 THEN 220
130   IF 12*X2 <  350000 THEN 230
140   IF 12*X2 <  500000 THEN 240
160   C2=4.00
165   GO TO 260
190   C2=6.00
195   GO TO 260
200   C2=5.60
205   GO TO 260
210   C2=5.20
215   GO TO 260
220   C2=5.00
225   GO TO 260
230   C2=4.45
235   GO TO 260
240   C2=4.25
```

```
260  IF C*12*X1+C2*12*X2 > 1250000 THEN 330
265  P3=(-.0009*X2**2+28.80*X2-C2*X2)*12
268  P1=(-.0005*X1**2+25*X1-C*X1)*12
269  P2=(12*X1*C)*.12
270  P4=(12*X2*C2)*.12
280  P=P1-P2+P3-P4
290  IF P > B THEN 310
300  GO TO 330
310  A1=12*X1
312  A2=12*X2
314  R1=C
316  R2=C2
317  K1=P1
318  K2=P2
319  K3=P3
320  K4=P4
321  B=P
330  NEXT X2
335  NEXT X1
340  PRINT A1,A2
345  PRINT B
350  PRINT R1,R2
355  PRINT K1,K2,K3,K4
360  STOP
370  END

200400          88800
3.90762E+06
4               5
2535060         96192.0      1522032      53280.0
```

We get x_1 = 200,400/12 = 16,700 per month and x_2 = 88,800/12 = 7400 per month to make and sell in order to maximize profit.

Let us now focus our search for 16,600 $\leqslant x_1 \leqslant$ 16,800 and 7300 $\leqslant x_2 \leqslant$ 7500 per month. After running the program with these bounds we get the true optimal solution, which is the company should borrow \$1,250,000 at 12% to buy 200,004 brass sections and 88,800 chrome fittings. They will pay \$4 and \$5 per unit, respectively, and price them at −.0005 x 200.004/12 + 25 = \$16.67 for the brass units, and −.009 x 88,800/12 + 28.80 = \$22.14 for the chrome units, and earn \$3,917,730 profit in the coming year.

It might also be worth investigating borrowing more money at a higher percent. In addition, more inventories and more price lists could be considered.

EXERCISES

7.1 The Black River Company has modeled its production at its Lake Superior plant to be $P = (-.22x_1 + 800)x_1 - 40x_1 - 900y_1 + (-.25x_2 + 600)x_2 - 30x_2 - 1800y_2 + (-.18x_3 + 900)x_3 - 35x_3 - 1500y_3$ subject to $x_1 \geqslant 0, x_2 \geqslant 0, x_3 \geqslant 0$, and $y_1 = 1$ if $x_1 > 0, y_2 = 1$ if $x_2 > 0, y_3 = 1$ if $x_3 > 0, y_1 = 0$ if $x_1 = 0, y_2 = 0$ if $x_2 = 0, y_3 = 0$ if $x_3 = 0$, and $9x_1 + 17x_2 \leqslant 70{,}000$ worker-hours in Department I and $3x_1 + 4x_2 + 16x_3 \leqslant 50{,}000$ worker-hours in Department II. (Note that fixed costs and unit costs are included.)

Find the maximum profit solution, including the prices Black River should charge in order to maximize profit (note the price-quantity curves in the profit function).

7.2 Substitute the optimal solution (x_1, x_2, and x_3) into the two worker constraints in exercise 7.1. If the two departments are not being fully utilized, consider reassigning workers from one department to another, etc. Should the company hire additional workers?

7.3 The Marketing Department at Black River estimates that an additional $5000 (added to their advertising fixed costs) for each of the three products in exercise 7.1 would change their price-quantity curves (advertising influencing demand) to $P_1 = -.22x_1 + 900$, $P_2 = -.24x_2 + 700$, and $P_3 = -.20x_3 + 1100$, respectively.

Rerun the program with these changes to see if the advertising campaign would be worth it.

7.4 The Inverness Tool Company wants to buy steel rods and wooden handles and make them into their Grade A screwdrivers. Their unit cost is $.20. The price-quantity curve for the screwdrivers is $P_1 = -.00001333x_1 + 3$. The price lists for the steel rods and handles are as follows:

Steel Rods (in one screwdriver lengths)	Unit price ($)
0 - 9,999	.35
10,000 - 39,999	.30
40,000 - 79,999	.25
80,000 or more	.21

Handles (one per screwdriver)	Unit price ($)
0 - 14,999	.18
15,000 - 35,999	.16
36,000 - 74,999	.15
75,000 or more	.14

The company will have to borrow the money at 11.75% in order to finance the project. How many units of steel rods and handles should they buy and make into screwdrivers and sell in order to maximize profit on this project?

7.5 The Old California Company makes and sells Product Z. One unit of Product Z consists of 1 unit of A, 3 units of B, and 2 units of C. The price-quantity lists for A, B, and C are as follows:

Supplier–Cardiff Ltd.
(1 unit needed per Z)

Number of units of A	Unit price ($)
0 - 99,999	1.00
100,000 - 499,999	.95
500,000 or more	.85

Supplier–Fort Frances Corp.
(3 units per Z)

Number of units of B	Unit price ($)
0 - 199,999	.27
200,000 - 599,999	.25
600,000 - 1,999,999	.23
2,000,000 or more	.20

Supplier–Copper Harbor
International Ltd.
(2 units per Z)

Number of units of C	Unit price ($)
0 - 499,999	.32
500,000 - 999,999	.30
1,000,000 - 1,499,999	.28
1,500,000 or more	.25

The assembly cost is $3 per unit, and the price-quantity curve is estimated at $P_1 = -.00001x_1 + 20$. The company has the money to purchase the units of A, B, and C outright. However, since that money could be invested in government bonds at 10%, the cost of the capital is 10%. Find the number of units of Product Z that the company should make and sell in order to maximize profit.

7.6 The Rogues Shirt and Pant Company will buy cotton cloth in quantity from Killarney Weavers Ltd. and make it into their rogue shirts and pants. It takes 2 meters of cloth to make a shirt and 3 meters to make a pair of pants. The proportion of different sizes to make has been figured out and will be adhered to. The price-quantity curve for shirts is estimated at $P_1 = -.000008x_1 + 32$, where x_1 is the number of shirts produced and P_1 is the selling price.

Shirts cost \$3.80 to make for the first 25,000 and \$5 (due to overtime) for any amount over 25,000. The fixed advertising cost for shirts is \$10,000. The price-quantity curve for pants is $P_2 = -.00001x_2 + 50$, where x_2 is the number of pairs and P_2 is the selling price. Pants cost \$5 to make regardless of the number made. \$12,000 will be spent on advertising the rogue line of pants.

One shirt and one pair of pants spend time in each of five departments according to the following schedule.

Time in minutes			
Shirt	Pair of pants	Department	Minutes available per year
3	5	Dyeing	1,000,000
4	4	Finishing	1,000,000
2	3	Printing	800,000
3	5	Cutting	800,000
7	8	Sewing	2,000,000

The company will borrow the money at 11% to buy the cotton from Killarney at the following price schedule:

Number of meters	Unit price (\$)
0 - 99,999	1.80
100,000 - 499,999	1.75
500,000 - 999,999	1.70
1,000,000 - 1,999,999	1.65
2,000,000 or more	1.60

The company has 5000 shirts and 4000 pairs of pants from last year's inventory. Find the level of production of shirts and pants and the prices to charge to maximize profit while selling all of last year's inventory and this year's too.

7.7 If necessary, redistribute workers in exercise 7.6 in order to make maximum use of the time in the five production departments. Increase profit through this reassignment process, if possible. If not, recommend layoffs.

7.8 Harrisonville Manufacturing makes and sells parsons tables. They buy the legs (4 per table) from Pacific Forestry, obtaining quantity discounts (see table below). They buy the table tops from Grand Marais Forest Products (price list below). The price-quantity curve for the finished tables is estimated at $P_1 = -.000025x_1 + 120$. The inplant construction cost is \$10 per table for the first 20,000 tables and \$15 for any tables past 20,000 units (due to overtime pay). Advertising will run \$5000 for the year. The company will borrow the money for advertising, the inplant cost, and the legs and table top inventory at 11.5%. Find the maximum profit solution to this problem.

Pacific Forestry table legs		Grand Marais Forest Products table tops	
Number	Unit cost per leg ($)	Number	Unit cost per table top ($)
0 - 49,999	2.50	0 - 9,999	11.00
50,000 - 99,999	2.25	10,000 - 29,999	10.00
100,000 or over	2.00	30,000 - 49,999	9.50
		50,000 or over	9.00

7.9 The Albertville Steel and Glass Company makes rotators which they sell to several different companies. The price-quantity curve for the rotators for the next year is estimated to be $P_1 = -.000001x_1 + 47$, where P_1 is the price and x_1 is the quantity sold. Rotators are composed of 2 steel clamps, 4 rubber disks, and 3 glass sheets (price list below). It takes 1 hour in Department 1 and .5 hours in Department 2 to make a rotator. There is a fixed cost of $4000 for advertising.

No. of steel clamps	Unit price($)	No. of rubber disks	Unit price($)	No. of glass sheets	Unit price($)
0 - 499,999	.40	0 - 999,999	.11	0 - 1,499,999	.21
500,000 - 1,499,999	.38	1,000,000 - 1,999,999	.09	1,500,000 or over	.15
1,500,000 or over	.35	2,000,000 or over	.07		

Albertville also makes power hand saws. The price quantity curve for the saws is $P_2 = -.00008x_2 + 66$, where P_2 is the price and x_2 is the quantity sold. The fixed advertising cost for the saws is $4500. The saws are composed of the metal housing, the blade, and the electrical workings. The price lists are below. It takes 2 hours in Department 1 and 1 hour in Department 2 to make one saw. There are 500,000 worker-hours available in Department 1 and 400,000 worker-hours available in Department 2. The unit assembly cost for rotators is $9. The unit assembly cost for saws is $11.

No. of metal housing w/blade	Unit price ($)	No. of elec. workings	Unit price ($)
0 - 49,999	3.90	0 - 74,999	3.60
50,000 - 99,999	3.60	75,000 - 159,999	3.40
100,000 or more	3.40	160,000 or more	3.25

The company will borrow money at 11% interest to finance these two operations. Find the maximum profit solution.

7.10 Redo exercise 7.9 assuming that 10,000 units of rotators and 7000 saws must be left in the year-end inventory. Then maximize profit.

7.11 Investigate changing the 500,000 and/or 400,000 worker-hour limits in exercise 7.9.

7.12 The Huron Houghton Company is trying to decide on the level of production and price for their Grade A plastic pipe for the next two years. Currently they have 20,000 feet of this pipe in their warehouse. They want to end this next year with 30,000 feet of pipe and to end the second year with 50,000 feet in inventory. The price-quantity curve for the first year is estimated at $P_1 = -.000005x_1 + 18$ with a fixed advertising cost of $5000. The price-quantity curve for the second year is estimated at $P_2 = -.000005x_2 + 23$ with a fixed advertising cost of $6000. The price list for this year and estimated list for next year are below:

1980		1981	
Feet of sheet pipe	cost per foot ($)	Feet of sheet pipe	cost per foot ($)
0 - 9999	2.00	0 - 9999	2.20
10,000 - 49,999	1.90	10,000 - 49,999	2.08
50,000 - 149,999	1.80	50,000 - 149,999	1.95
150,000 - 599,999	1.70	150,000 - 599,999	1.80
600,000 or more	1.60	600,000 or more	1.70

The company is going to borrow the money at 11% this year and probably 12% next year to finance the purchase of the sheet pipe, which will be made into Grade A plastic pipe at a unit cost of $1.25 the first year and $1.40 the next year. The storage cost is negligible. Find the maximum profit solution. When should the company borrow and how much? What price should they charge each year and how much will they sell each year?

Suggested Reading

Bierman, Bonini & Hausman. *Quantitative Analysis for Business Decisions.* 5th ed. Homewood, IL.: Richard D. Irwin, 1977.

Charnes, A. and Cooper, W.W. *Management Models and Industrial Applications of Linear Programming*. New York: John Wiley & Sons, 1961.

Conley, William. *Pricing a Product.* International Journal of Mathematical Education in Science and Technology. 12 No. 1 (1981) pp 63-67.

Hillier, F.S. and Lieberman, G.J. *Introduction to Operations Research.* San Francisco: Holden-Day, 1967.

Jaaskelainen, Veikko. *Linear Programming and Budgeting.* New York: Petrocelli/Charter, 1975.

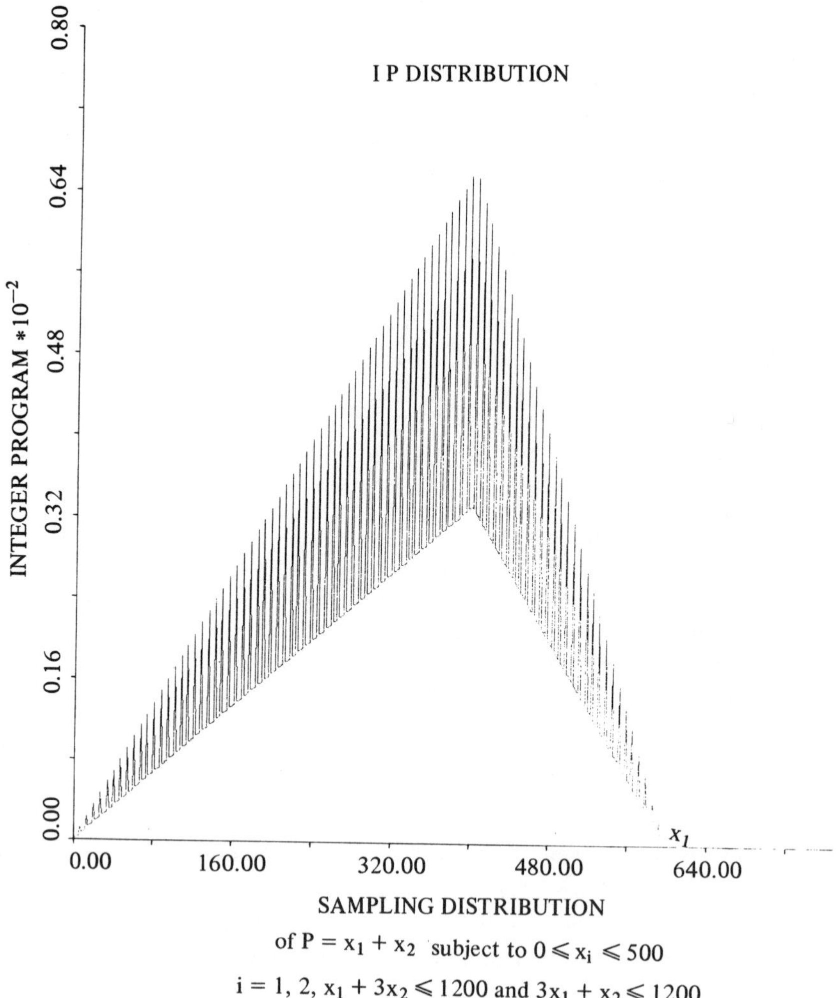

FIGURE 8.1 is the graph of all the answers of an integer programming problem that has constraints.

CHAPTER 8

More Business Applications

The success of Monte Carlo optimization makes it possible for business managers to write accurate nonlinear planning models that include fixed costs, discounts for buying in quantity, cost of capital, economic order factors, and accurate constraints. This chapter offers a few examples and exercises as a further illustration of the various Monte Carlo solution techniques.

EXAMPLE 8.1

The Northern Lights Company has to place orders for three items, A_1, A_2, and A_3, to be carried in their inventory. Demand is fairly constant for all three items so the only variables involved are the reorder quantities q_1, q_2, q_3 (the number of each item that the company requests each time they reorder).

The total demand for A_1, A_2, and A_3 is 10,000 units, 20,000 units, and 15,000 units, respectively. The cost of placing one order for A_1, A_2, and A_3 is O_1 = \$50, O_2 = \$100, and O_3 = \$75, respectively. The holding cost of one unit of A_1, A_2, and A_3 per time period is H_1 = \$2, H_2 = \$3, and H_3 = \$1.80, respectively.

Using the fairly standard inventory (EOQ) model, Cost = $H_i q_i/2 + O_i r_i/q_i$, we could write three individual programs and find the q_i's that minimize the inventory cost to be approximately $q_1 = 707$, $q_2 = 1154$, and $q_3 = 1118$.

However, one problem complicates the model. The first orders of the three products will all arrive at once, and they must be stored in a 5000 cubic meter warehouse. One unit of A_1 occupies 3 cubic meters, one unit of A_2 occupies 1 cubic meter, and one unit of A_3 occupies 6 cubic meters. Therefore, the problem becomes to find q_1, q_2, and q_3 such that we minimize $C = H_1q_1/2 + O_1r_1/q_1 + H_2q_2/2 + O_2r_2/q_2 + H_3q_3/2 + O_3r_3/q_3$ subject to $3q_1 + q_2 + 6q_3 \leqslant 5{,}000$.

The program follows:

```
5     REM FIND REORDER QUANTITIES TO MINIMIZE
10    REM INVENTORY COSTS SUBJECT TO
15    REM WARE HOUSE CONSTRAINTS
20    X=1
30    B=999999
40    FOR I=1 TO 20000
50    Q3=INT(RND(X)*833)+1
60    Q2=INT(RND(X)*(5000-6*Q3))+1
70    Q1=INT(RND(X)*((5000-6*Q3-Q2))+1
80    C1=2*Q1/2.+50*10000/Q1
90    C2=3*Q2/2+100*20000/Q2
100   C3=1.80*Q3/2+75*15000/Q3
110   C=C1+C2+C3
120   IF C < B THEN 140
130   GO TO 180
140   B=C
150   A1=Q1
160   A2=Q2
170   A3=Q3
180   NEXT I
190   PRINT A1,A2,A3
200   PRINT B
210   STOP
220   END

409             1077            444
8037.39

436             1075            436
8028.43

420             1082            442
8024.95

416             1087            444
8021.73
```

```
367             1106            464
8038.89

409             1078            444
8037.17

436             1075            436
8028.43

420             1082            442
8024.95

416             1087            444
8021.73

367             1106            464
8038.89

409             1078            444
8037.17
```

After looking at the printouts of the runs of 20,000, we decide to focus Q_3 between 350 and 450. Line 50 is changed:

```
50    Q3=INT(RND(X)*101+350

431             1004            446
8012.95

438             1003            446
8001.89

444              981            445
8008.95

439             1004            446
7999.81
```

A further focus yields:

```
5     REM FIND REORDER QUANTITIES TO MINIMIZE
10    REM INVENTORY COSTS SUBJECT TO
15    REM WAREHOUSE CONSTRAINTS
20    X=1
```

```
30    B=999999
40    FOR Q1=350 TO 450 STEP 2
50    FOR Q2=1000 TO 1100 STEP 2
60    FOR Q3=440 TO 470
70    IF 3*Q1+Q2+6*Q3 > 5000 THEN 180
80    C1=2*Q1/2.+50*10000/Q1
90    C2=3*Q2/2+100*20000/Q2
100   C3=1.80*Q3/2+75*15000/Q3
110   C=C1+C2+C3
120   IF C < B THEN 140
130   GO TO 180
140   B=C
150   A1=Q1
160   A2=Q2
170   A3=Q3
180   NEXT Q3
181   NEXT Q2
182   NEXT Q1
190   PRINT A1,A2,A3
200   PRINT B
210   STOP
220   END

400        1004          466
7981.59
```

Finally we search all the points in the region $390 \leqslant q_1 \leqslant 410$, $990 \leqslant q_2 \leqslant 1010$, and $460 \leqslant q_3 \leqslant 470$:

```
5     REM FIND REORDER QUANTITIES TO MINIMIZE
10    REM INVENTORY COSTS SUBJECT TO
15    REM WAREHOUSE CONSTRAINTS
20    X=1
30    B=999999
40    FOR Q1=390 TO 410
50    FOR Q2=990 TO 1010
60    FOR Q3=460 TO 470
70    IF 3*Q1+Q2+6*Q3 > 5000 THEN 180
80    C1=2*Q1/2.+50*10000/Q1
90    C2=3*Q2/2+100*20000/Q2
100   C3=1.80*Q3/2+75*15000/Q3
110   C=C1+C2+C3
120   IF C < B THEN 140
130   GO TO 180
140   B=C
```

```
150  A1=Q1
160  A2=Q2
170  A3=A3
180  NEXT Q3
181  NEXT Q2
182  NEXT Q1
190  PRINT A1,A2,A3
200  PRINT B
210  STOP
220  END

400                 992                468
7979.18
```

Therefore, Northern Lights should reorder 400 units of A_1, 992 units of A_2, and 468 units of A_3 to achieve a minimum cost of \$7979.18 subject to the warehouse constraint.

EXAMPLE 8.2

The Neuchatel Tea Company makes three kinds of tea: International Best (selling for \$5 a pound), Orange Smooth (selling for \$4.20 a pound), and Popular Blend (selling for \$3.50 a pound). They also make the one-pound size cans for the teas. Each of the three teas has its own label stamped on the standard can. Therefore, a fixed cost for putting on and changing the label stamp (the dyes must be removed, etc.) for each production run is incurred. The fixed cost is \$150 for International Best, \$90 for Orange Smooth, and \$80 for Popular Blend.

The teas are blended from Indian tea (\$2.50 per pound and 3000 pounds available), Sri Lanka tea (\$2.75 per pound and 2500 pounds available), and Indonesian tea (\$2.95 per pound and 2000 pounds available).

The recipe for International Best calls for exactly 50% Indonesian tea, 30% Sri Lanka tea, and 20% Indian tea. Orange Smooth must be at least 60% Sri Lanka and Indonesian tea. Popular Blend must be at least 30% Indian tea.

Let x_1, y_1, and z_1 equal the number of pounds of Indian tea blended into International Best, Orange Smooth, and Popular Blend, respectively. Let x_2, y_2, and z_2 equal the number of pounds of Sri Lanka tea blended into International Best, Orange Smooth, and Popular Blend, respectively. Let x_3, y_3, and z_3 equal the number of pounds of Indonesian tea blended into International Best, Orange Smooth, and Popular Blend, respectively.

Therefore, the problem can be stated as maximize $P = P1 + P2 + P3$, where $P1 = 2.50x_1 + 2.25x_2 + 2.05x_3 - 150F_1$, $P2 = 1.70y_1 + 1.45y_2 + 1.25y_3 - 90F_2$, and $P3 = 1.00z_1 + .75z_2 + .55z_3 - 80F_3$, where

$$F_1 = 1 \text{ if } x_1 > 0 \text{ or } x_2 > 0 \text{ or } x_3 > 0$$
$$F_1 = 0 \text{ if } x_1 = x_2 = x_3 = 0$$
$$F_2 = 1 \text{ if } y_1 > 0 \text{ or } y_2 > 0 \text{ or } y_3 > 0$$
$$F_2 = 0 \text{ if } y_1 = y_2 = y_3 = 0$$
$$F_3 = 1 \text{ if } z_1 > 0 \text{ or } z_2 > 0 \text{ or } z_3 > 0$$
$$F_3 = 0 \text{ if } z_1 = z_2 = z_3 = 0$$

subject to

$$x_1 + y_1 + z_1 \leqslant 3000$$
$$x_2 + y_2 + z_2 \leqslant 2500$$
$$x_3 + y_3 + z_3 \leqslant 2000$$

and

$$y_2 + y_3 \geqslant .6\,(y_1 + y_2 + y_3)$$
$$z_1 \geqslant .3\,(z_1 + z_2 + z_3)$$

and

$$.20(x_1 + x_2 + x_3) = x_1$$
$$.30(x_1 + x_2 + x_3) = x_2$$
$$.50(x_1 + x_2 + x_3) = x_3$$

Let us start work on this problem first by rearranging the system of three equations and three unknowns to:

$$x_1 + x_2 + x_3 = 5x_1$$
$$3x_1 + 3x_2 + 3x_3 = 10x_2$$
$$x_1 + x_2 + x_3 = 2x_3$$

and then to:

$$x_1 + x_2 - x_3 = 0$$
$$3x_1 - 7x_2 + 3x_3 = 0$$
$$-4x_1 + x_2 + x_3 = 0$$

Then we solve this system of equations using the diagonal technique as follows:

$$\begin{array}{rrrr}
1 & 1 & -1 & 0 \\
3 & -7 & 3 & 0 \\
-4 & 1 & 1 & 0 \\
1 & 1 & -1 & 0 \\
0 & -10 & 6 & 0 \\
0 & 5 & -3 & 0 \\
1 & 1 & -1 & 0 \\
0 & 1 & -3/5 & 0 \\
0 & 5 & -3 & 0 \\
1 & 0 & -2/5 & 0 \\
0 & 1 & -3/5 & 0 \\
0 & 5 & -3 & 0 \\
1 & 0 & -2/5 & 0 \\
0 & 1 & -3/5 & 0 \\
0 & 0 & 0 & 0
\end{array}$$

For arbitrary x_3 from 0 to 2000, $x_1 = 2/5x_3$ and $x_2 = 3/5x_3$.

Now our strategy will be to read in a random number between 0 and 2000 for x_3 and then use this to determine x_1 and x_2. Then we will read in random numbers for y_1, y_2, and y_3, bounding them by $3000 - x_1$, $2500 - x_2$, and $2000 - x_3$, respectively. Then we will read in random numbers for z_1, z_2, and z_3, bounding them by $3000 - x_1 - y_1$, $2500 - x_2 - y_2$, and $2000 - x_3 - y_3$, respectively. Then we will check these numbers in the remaining constraints and proceed with the function.

We will do this 20,000 times for a start. The program follows:

```
4       REM BLENDING PROBLEM TO
5       REM MAXIMIZE PROFIT MADE
6       REM NONLINEAR BY FIXED COSTS
10      B=-999999
12      N=0
14      X=1
15      FOR I=1 TO 20000
```

```
17   X3=INT(RND(X)*2001)
18   X1=.4*X3
19   X2=.6*X3
20   Y1=INT(RND(X)*(3001-X1))
21   Y2=INT(RND(X)*(2501-X2))
22   Y3=INT(RND(X)*(2001-X3))
23   Z1=INT(RND(X)*(3001-X1-Y1))
24   Z2=INT(RND(X)*(2501-X2-Y2))
25   Z3=INT(RND(X)*(2001-X3-Y3))
30   IF Y2+Y3 <  .6*(Y1+Y2+Y3) THEN 260
35   IF Z1  < .3*(Z1+Z2+Z3) THEN 260
40   N=N+1
41   P1=2.50*X1+2.25*X2+2.05*X3-150
42   P2=1.70*Y1+1.45*Y2+ 1.25*Y3-90
43   P3=1.00*Z1+.75*Z2+.55*Z3-80
44   P=P1+P2+P3
45   IF X3=0 THEN 100
50   IF Y2=0 THEN 60
55   GO TO 80
60   IF Y3=0 THEN 64
62   GO TO 80
64   IF Y1=0 THEN 110
80   IF Z1=0 THEN 82
81   GO TO 140
82   IF Z2=0 THEN 84
83   GO TO 140
84   IF Z3=0 THEN 120
90   GO TO 140
100  P=P+150
105  GO TO 50
110  P=P+90
115  GO TO 80
120  P=P+80
140  IF P > B THEN 160
150  GO TO 260
160  A1=X1
170  A2=X2
180  A3=X3
190  A4=Y1
200  A5=Y2
210  A6=Y3
220  A7=Z1
230  A8=Z2
240  A9=Z3
250  B=P
260  NEXT I
265  PRINT N
270  PRINT A1,A2,A3
280  PRINT A4,A5,A6
```

```
290  PRINT A7,A8,A9
300  PRINT B
310  STOP
320  END
6235
787.600         1181.40         1969
659             1222            28
1468            72              3
12794.4
```

After several runs like this, it appears that z_3 should be set equal to zero. This adjustment yields the following printouts:

```
6788
740.000  1110     1850
871      1377     71
1291     4        0
12680.1
6716
795.200  1192.80  1988
702      1267     12
1420     7        0
12898.0
6664
681.600  1022.40  1704
1029     1476     295
1277     2        0
12714.3
```

From this it appears that z_2 should be set equal to zero. We do this and rerun the program. The best printout produces:

```
4818
796.800  1195.20  1992
714      1285     0
1469     0        0
12990.8
```

A focus search could produce the true optimum, but this solution is very nearly optimal. Therefore, the Neuchatel Tea Company should blend 796.8 pounds, 714 pounds, and 1469 pounds of Indian tea into International Best, Orange Smooth, and Popular Blend, respectively. They should also blend 1195.20 pounds and 1285 pounds of Sri Lanka tea into International Best and Orange Smooth, respectively. Finally, they should blend 1992 pounds of Indonesian tea into International Best Tea.

EXAMPLE 8.3

A space shuttle has a 48 cubic meter compartment in which nonprofit scientific research projects are to be put. The shuttle screening committee has fifteen project applications for the first flight. The committee is under considerable pressure to make maximum use of the shuttle; therefore, they have decided to do a cargo loading problem (maximize the total importance rating) for each flight. All projects left out from the first flight will be considered for the next flight along with new applications submitted by the new filing date. The compartment has a weight limit of 8 metric tons (8000 kilograms). Also, projects G and H cannot go together because one would ruin the other. And if Project A goes, B must go with it, but B can go without A.

Below is a listing of the projects, their weight, volume, and importance rating as judged by the committee:

Space Shuttle First Flight

Project	Weight (tons)	Volume (cubic meters)	Importance rating
A	.25	1	20
B	.2	1	10
C	.3	6	90
D	.4	5	58
E	.5	8	70
F	.52	2	14
G	1.2	9	92
H	2.0	6	85
I	1.6	4	60
J	.2	1	9
K	.8	7	54
L	.7	7	106
M	.4	6	45
N	.5	5	59
O	.2	6	75

Therefore, our problem is to maximize $P = 20A + 10B + 90C + 58D + 70E + 14F + 92G + 85H + 60I + 9J + 54K + 106L + 45M + 59N + 75O$ (letter O) subject to all variables are 0 or 1, and $G + H \leqslant 1, A \leqslant B$, $.25A + .2B + .3C + .4D + .5E + .52F + 1.2G + 2.0H + 1.6I + .2J + .8K + .7L + .4M + .5N + .2O \leqslant 8$ metric tons, and $A + B + 6C + 5D + 8E + 2F + 9G + 6H + 4I + J + 7K + 7L + 6M + 5N + 6O \leqslant 48$ cubic meters. The program follows:

```
4    REM CARGO LOADING PROBLEM.
5    REM TRYING TO MAXIMIZE PAYOFF
6    REM OF SCIENTIFIC PROJECTS ON
7    REM A SINGLE FLIGHT OF THE
8    REM SPACE SHUTTLE, SUBJECT TO
9    REM VOLUME AND WEIGHT CONSTRAINTS
10   B9=-999999
20   FOR A=0 TO 1
30   FOR B=0 TO 1
40   FOR C=0 TO 1
50   FOR D=0 TO 1
60   FOR E=0 TO 1
70   FOR F=0 TO 1
80   FOR G=0 TO 1
90   FOR H=0 TO 1
100  FOR I=0 TO 1
110  FOR J=0 TO 1
120  FOR K=0 TO 1
130  FOR L=0 TO 1
140  FOR M=0 TO 1
150  FOR N=0 TO 1
160  FOR O=0 TO 1
170  IF G+H > 1 THEN 470
180  IF A > B THEN 470
190  X1=.25*A+.2*B+.3*C+.4*D+.5*E
200  X2=.52*F+1.2*G+2.0*H+1.6*I+.2*J
210  X3=.8*K+.7*L+.4*M+.5*N+.2*O
220  IF X1+X2+X3 > 8 THEN 470
230  Y1=A+B+6*C+5*D+8*E
240  Y2=2*F+9*G+6*H+4*I+J
250  Y3=7*K+7*L+6*M+5*N+6*O
260  IF Y1+Y2+Y3 >  48 THEN 470
270  P1=20*A+10*B+90*C+58*D+70*E
280  P2=14*F+92*G+85*H+60*I+9*J
290  P3=54*K+106*L+45*M+59*N+75*O
300  P=P1+P2+P3
310  IF P  >  B9 THEN 330
320  GO TO 470
330  A1=A
340  B1=B
```

```
350  C1=C
360  D1=D
370  E1=E
380  F1=F
390  G1=G
400  H1=H
410  I1=I
420  K1=K
430  L1=L
440  M1=M
450  N1=N
460  O1=O
465  B9=P
470  NEXT O
480  NEXT N
490  NEXT M
500  NEXT L
510  NEXT K
520  NEXT J
530  NEXT I
540  NEXT H
550  NEXT G
560  NEXT F
570  NEXT E
580  NEXT D
590  NEXT C
600  NEXT B
610  NEXT A
620  PRINT A1,B1,C1,D1,E1
630  PRINT F1,G1,H1,I1,J1
640  PRINT K1,L1,M1,N1,O1
650  PRINT B9
660  STOP
670  END

1      1      1     1    0
0      0      1     1    0
1      1      0     1    1
617
```

Therefore, projects A, B, C, D, H, I, K, L, N, and O should go on the first flight in order to maximize the total importance rating, subject to the volume and weight constraints.

The space shuttle also has a 44 cubic meter compartment in which industrial projects submitted by private companies can be placed. The maximum allowable weight in this compartment is 7 metric tons (7000 kilograms). The committee has been authorized to charge \$25,000 per cubic meter for project space. Therefore, which of the twelve project applications listed below with their weight and volume requirements should be selected in order to maximize profit, subject to the constraints?

Space Shuttle First Flight

Project	Weight (tons)	Volume (cubic meters)
A	.9	5
B	2.0	9
C	1.4	8
D	.4	3
E	.4	2
F	.88	6
G	.4	3
H	2.2	11
I	1.6	6
J	1.2	2
K	3.1	5
L	1.4	8

The problem is maximize $P = 25{,}000(5A + 9B + 8C + 3D + 2E + 6F + 3G + 11H + 6I + 2J + 5K + 8L)$ subject to all variables have the values 0 or 1, $.9A + 2.0B + 1.4C + .4D + .4E + .88F + .4G + 2.2H + 1.6I + 1.2J + 3.1K + 1.4L \leq 7$ tons, and $5A + 9B + 8C + 3D + 2E + 6F + 3G + 11H + 6I + 2J + 5K + 8L \leq 44$ cubic meters.

The program to maximize the profit is as follows:

```
5     REM TRYING TO MAXIMIZE PROFIT
6     REM FOR INDUSTRIAL PROJECTS ON
7     REM A SINGLE FLIGHT OF THE
8     REM SPACE SHUTTLE, SUBJECT TO
9     REM VOLUME AND WEIGHT CONSTRAINTS
10    B9=-999999
20    FOR A=0 TO 1
30    FOR B=0 TO 1
40    FOR C=0 TO 1
50    FOR D=0 TO 1
60    FOR E=0 TO 1
70    FOR F=0 TO 1
80    FOR G=0 TO 1
90    FOR H=0 TO 1
100   FOR I=0 TO 1
```

```
110   FOR J=0 TO 1
120   FOR K=0 TO 1
130   FOR L=0 TO 1
140   X1=.9*A+2.0*B+1.4*C+.4*D+.4*E+.88*F
150   X2=.4*G+2.2*H+1.6*I+1.2*J+3.1*K+1.4*L
160   IF X1+X2 > 7 THEN 370
170   Y1=5*A+9*B+8*C+3*D+2*E+6*F
180   Y2=3*G+11*H+6*I+2*J+5*K+8*L
190   IF Y1+Y2 > 44 THEN 370
200   P1=5*A+9*B+8*C+3*D+2*E+6*F
210   P2=3*G+11*H+6*I+2*J+5*K+8*L
220   P=25000*(P1+P2)
230   IF P > B9 THEN 250
240   GO TO 370
250   A1=A
260   B1=B
270   C1=C
280   D1=D
290   E1=E
300   F1=F
310   G1=G
320   H1=H
330   I1=I
340   J1=J
350   K1=K
360   L1=L
365   B9=P
370   NEXT L
380   NEXT K
390   NEXT J
400   NEXT I
410   NEXT H
420   NEXT G
430   NEXT F
440   NEXT E
450   NEXT D
460   NEXT C
470   NEXT B
480   NEXT A
490   PRINT A1,B1,C1,D1
500   PRINT E1,F1,G1,H1
510   PRINT I1,J1,K1,L1
520   PRINT B9
530   STOP
540   END

0             0             1             1
0             1             1             1
0             0             0             1
975000
```

Therefore, industrial projects C, D, F, G, H, and L should be sent on the first flight to yield maximum revenue of $975,000.

EXAMPLE 8.4

The North Bay Steel Supply Company wants to buy some Grade C steel beams. The holding cost (average storage cost per unit per year) is 18% of the price. On the average, $Q/2$ units are in the warehouse at any given time. Total demand for the year is 15,000 units. The reorder cost for each order is $40. The cost function to be minimized is $C = .18QB/2 + 40\,(15{,}000/Q) + 15{,}000B$, where Q is the number of units that are ordered for each reorder and B is the unit price.*

However, the supplier gives discounts for buying in quantity as follows:

Number of Units	Unit price ($)
0 - 2,499	20.00
2,500 - 4,999	17.50
5,000 - 9,999	15.00
10,000 or more	13.50

The program written to minimize C is below:

```
10   REM INVENTORY REORDER QUANTITY Q
20   REM TO MINIMIZE COST GIVEN DISCOUNTS
30   REM FOR BUYING IN QUANTITY
40   B1=999999
50   FOR Q=1 TO 15000
60   IF Q < 2500 THEN 100
70   IF Q < 5000 THEN 110
80   IF Q < 10000 THEN 120
90   GO TO 130
100  B=20.00
105  GO TO 140
110  B=17.50
115  GO TO 140
120  B=15.00
125  GO TO 140
130  B=13.50
140  C=.18*Q*B/2.+40*15000/Q+15000*B
150  IF C < B1 THEN 170
160  GO TO 190
170  Q1=Q
```

*This cost equation is similar to the one used by Thomas Cook and Robert Russell in *Introduction to Management Science* (Englewood Cliffs, N.J.: Prentice-Hall, 1977), p. 396

```
180  B1=B
190  NEXT Q
200  PRINT Q1
210  PRINT B1
220  STOP
230  END

10000
214710
```

Therefore, the company should order 10,000 Grade C beams every eight months to minimize its inventory costs.

EXAMPLE 8.5

Southern Cross Limited wants to invest units of $1 million in some or all of the following investments for a year:

	Yield(%)	Risk rating	Liquidity rate	Possible number of units for investment
Tin mine	22	8	4	0,1 or 2
Silver mine	35	8	4	0,1 or 2
Coal mine	25	6	2	0,1 or 2
Oil field	30	10	4	0,1 or 2
Coffee plantation	15	3	3	0,1 or 2
Cattle ranch	11	2	3	0,1,2, or 3
Sugar processing plant	10	3	2	0,1,2, or 3
Steel company	12	2	5	0,1,2, or 3
Shipping company	20	5	2	0,1,2, or 3
Savings account	6	1	1	0,1,2, or 3

The investment department has estimated the yields, risk ratings, and liquidity ratings. Also, management has decreed that no more than $2 or $3 million will be invested in any one project (as specified in the right-hand column of the chart). In addition, average risk is to be no more than four. Average liquidity is to be less than or equal to 2.5 or at least five different investments must be made (diversity condition). So only the diversity or liquidity condition must hold (risk must hold).

Therefore, Southern Cross must maximize $P = .22x_1 + .35x_2 + .25x_3 + .30x_4 + .15x_5 + .11x_6 + .10x_7 + .12x_8 + .20x_9 + .06x_0$ subject to all $x_i = 0, 1$, or 2 (or 3 for x_i where $i = 6, 7, 8, 9, 10$), $(8x_1 + 8x_2 + 6x_3 + 10x_4 + 3x_5 + 2x_6 + 3x_7 + 2x_8 + 5x_9 + x_0)/10 \leqslant 4$, and at least five $x_i > 0$ or $(4x_1 + 4x_2 + 2x_3 + 4x_4 + 3x_5 + 3x_6 + 2x_7 + 5x_8 + 2x_9 + x_0)/10 \leqslant 2.5$.

The program follows:

```
4     REM TEN VARIABLE FINANCIAL INVESTMENT
5     REM PROBLEM TO MAXIMIZE YIELD
6     REM SUBJECT TO RISK, LIQUIDITY
7     REM DIVERSITY CONSTRAINTS
8     REM BUY ONLY DIVERSITY
9     REM OR LIQUIDITY CONSTRAINT HAS
10    REM TO HOLD, NOT BOTH.
11    REM HOWEVER RISK MUST HOLD.
20    B=-999999
30    N1=0
40    FOR X1=0 TO 2
50    FOR X2=0 TO 2
60    FOR X3=0 TO 2
70    FOR X4=0 TO 2
80    FOR X5=0 TO 2
90    FOR X6=0 TO 3
100   FOR X7=0 TO 3
110   FOR X8=0 TO 3
120   FOR X9=0 TO 3
130   FOR X0=0 TO 3
134   R1=8*X1+8*X2+6*X3+10*X4+3*X5
135   R2=2*X6+3*X7+2*X8+5*X9+X0
136   R=R1+R2
137   R3=R/10.
138   IF R3 > 4 THEN 490
140   N=0
150   IF X1=0 THEN 160
155   N=N+1
160   IF X2=0 THEN 170
165   N=N+1
170   IF X3=0 THEN 180
175   N=N+1
180   IF X4=0 THEN 190
185   N=N+1
190   IF X5=0 THEN 200
195   N=N+1
200   IF X6=0 THEN 210
```

```
205  N=N+1
210  IF X7=0 THEN 220
215  N=N+1
220  IF X8=0 THEN 230
225  N=N+1
230  IF X9=0 THEN 240
235  N=N+1
240  IF X0=0 THEN 250
245  N=N+1
250  IF N > 4 THEN 320
255  GO TO 290
290  L1=4*X1+4*X2+2*X3+4*X4+3*X5
291  L2=3*X6+2*X7+5*X8+2*X9+X0
292  L=L1+L2
300  L3=L/10.
310  IF L3 > 2.5 THEN 490
320  N1=N1+1
330  P1=.22*X1+.35*X2+.25*X3+.30*X4+.15*X5
340  P2=.11*X6+.10*X7+.12*X8+.20*X9+.06*X0
350  P=P1+P2
360  IF P > B THEN 380
370  GO TO 490
380  A1=X1
390  A2=X2
400  A3=X3
410  X4=X4
420  A5=X5
430  Z6=X6
440  A7=X7
450  A8=X8
460  A9=X9
470  A0=X0
480  B=P
490  NEXT X0
500  NEXT X9
510  NEXT X8
520  NEXT X7
530  NEXT X6
540  NEXT X5
550  NEXT X4
560  NEXT X3
570  NEXT X2
580  NEXT X1
590  PRINT A1,A2,A3,A4,A5
600  PRINT A6,A7,A8,A9,A0
610  PRINT B
620  PRINT N1
630  STOP
640  END
```

```
0      2      0      0      2
3      1      3      0      3
1.97000
45336
```

Notice that lines 134, 135, 136, 137, and 138 check to see that the risk constraint holds. Also, lines 140 through 250 check the diversity constraint and lines 290 through 310 check the liquidity constraint. However, the logic is arranged so that only the diversity or the liquidity constraint must hold (the risk constraint is required from the problem statement).

The optimal solution (from the printout) is to invest \$2 million in the silver mine, \$2 million in the coffee plantation, \$3 million in the cattle ranch, \$1 million in the sugar processing plant, and \$3 million in both the steel company and the savings account to yield a maximum profit of \$1,970,000 for the year. This is the optimal solution out of the 45,336 solutions that satisfied the risk constraint and one or the other of the liquidity and diversity constraints.

It would be very instructive to rerun the program to print out all the answers that are nearly optimum (say, greater than 1.95 million, for example) and/or relax some of the constraints to see how much profit would improve.

Further, Southern Cross Limited has a production planning problem. They are making two products (A and B) in their western plant. Let x_1 be the number of units of A produced there per year and let x_2 be the number of units of B produced there per year. The unit costs are \$2 for A and \$2.40 for B. Fixed costs are \$10,000 for A and \$12,000 for B. Demand for A can be expressed as $y = -.0025x_1 + 100$, where y is the price and x_1 the resulting number of units of A sold at that price level. $Y = -.0020x_2 + 120$ expresses demand for B. Also, A and B are products that are finished at Southern Cross after purchasing the rough stock elsewhere. The inventory (standard EOQ) model for A is $100Q_1/2 + 50 x_1/Q_1$, where x_1 is the total demand and $Q1$ is the reorder quantity. Similarly, $80Q_1/2 + 35x_2/Q_2$ is the inventory cost equation for B.

The warehouse only has room for 80,000 units of B stock, and A stock is twice as large as B stock because two A stock units are welded together to make one unit of A. There are 100,000 worker-hours available; each unit of A needs three hours and B needs 2.5 hours. There are 120,000 hours of machine time available; A needs 1.9 hours per unit and B needs 3 hours. The problem then is to maximize $P = -.0025x_1{}^2 + 100x_1 - 10{,}000y_1 - (100Q_1/2 + 25(2x_1)/Q_1 - 2x_1 - .0020x_2{}^2 + 120x_2 - 1200y_2 - (80Q_2/2 + 35x_2/Q_2) - 2.4x_2$ subject to $x_1 \geqslant 0$, $x_2 \geqslant 0$, $y_1 = 1$ if $x_1 > 0$, $y_2 = 1$ if $x_2 > 0$, $y_1 = 0$ if $x_1 = 0$ and $y_2 = 0$ if $x_2 = 0$, $2x_1 + x_2 \leqslant 80{,}000$ warehouse capacity, $3x_1 + 2.5y_2 \leqslant 100{,}000$ worker-hours, $1.9x_1 + 3x_2 \leqslant 120{,}000$ hours of machine time, and $1 \leqslant Q_1 \leqslant 2x_1$ and $1 \leqslant Q_2 \leqslant x_2$. It is implied that if x_1 or $x_2 = 0$ then the corresponding inventory equation is dropped (preventing division by zero). The program follows:

```
5     REM TWO PRODUCT PRODUCT-MIX PROBLEM WITH
6     REM DEMAND CURVES, FIXED COSTS, INVENTORY
7     REM COSTS, UNIT COSTS, WAREHOUSE CONSTRAINTS.
8     REM AND WORKER AND MACHINE TIME LIMITS.
10    REM WE SOLVE BY DOING A SIMULATION.
11    REM WE ALSO OBTAIN REORDER QUANTITIES FOR
12    REM SUB ASSEMBLY PARTS.
20    B=-999999
25    N=0
30    X=1
40    FOR I=1 to 40000
50    X1=INT(RND(X)*33334)
60    Q1=INT(RND(X)*2*X1)+1
70    X2=INT(RND(X)*40001)
80    Q2=INT(RND(X)*X2)+1
90    IF 3*X1+2.5*X2 > 100000 THEN 300
100   IF 1.9*X1+3*X2 > 120000 THEN 300
110   IF 2*X1+X2 > 80000 THEN 300
120   P1=-.0025*X1**2+100*X1-10000-2*X1
130   P2=-.0020*X2**2+120*X2-12000-2.40*X2
140   I1=100*Q1/2.+25*2*X1/Q1
150   I2=80*Q2/2.+35*X2/Q2
160   P=P1-I1+P2-I2
161   N=N+1
170   IF X1=0 THEN 210
180   IF X2=0 THEN 230
190   IF P > B THEN 250
200   GO TO 300
210   P=P+10000
220   GO TO 180
230   P=P+12000
240   GO TO 190
250   A1=X1
260   B1=Q1
270   A2=X2
280   B2=Q2
```

```
290  B=P
300  NEXT I
305  PRINT N
310  PRINT A1,B1
320  PRINT A2,B2
330  PRINT B
340  STOP
350  END

20066
13608         1081
22562         607
2.40358E+06
```

Several runs like this one indicated that x_1 was between 13,700 and 13,900 and x_2 was between 22,800 and 23,000. There the simulation was narrowed accordingly and rerun a few times producing:

```
40000
13860         66
22968         187
2.47643E + 06
```

This and other runs put x_1 between 13,800 and 13,950 and x_2 between 22,900 and 23,050. All these points were checked nine times (with simulated Q_1's and Q_2's each time) and the best answer was:

```
22801
13885         161
23046         200
2.48033E + 06
```

Therefore, production of x_1 = 13,885 units of A (with reorders of A stock of 161 per order), and production of x_2 = 23,046 units of B (with reorders of B stock of 200 per order) should yield a nearly maximal profit of $2,480,330 per year.

EXAMPLE 8.6

The Saint Gervais Paint Company has a fixed cost and unit cost blending problem. They are considering making two kinds of paint, Red 1 and Red 2. Both paints are blends of four different ingredients as listed in the following chart.

Ingredients	Cost of fireproofing chemical per amount added per liter ($)	Amount used for Red 1 (liters)	Amt. avail. (liters)	Amt. used for Red 2 (liters)	Cost per liter ($)
Red pigment	3.00	x_1	1000	y_1	3.00
A varnish	3.00	x_2	900	y_2	4.00
B thinner	3.00	x_3	800	y_3	5.00
Type D5 driers	3.00	x_4	700	y_4	4.50

At least 5% of Red 1 must be red pigment. Red 1 must have exactly 20% B thinner and 20% type D5 driers in it. However, Red 1 is a fireproof paint and must have the fireproofing chemicals added to it; there is no appreciable increase in volume from the fireproofing. Red 2 must have 25% of all four ingredients. The company will sell Red 1 at $12 per liter and Red 2 at $9 per liter. Also, the fixed costs are judged to be $300 for Red 1 and $180 for Red 2. These fixed costs consist of advertising, washing out the mixer, and new labels for the paint cans.

Therefore, the problem is to maximize $P = 12(x_1 + x_2 + x_3 + x_4) - 6x_1 - 7x_2 - 8x_3 - 7.5x_4 - 300z_1 + 9(y_1 + y_2 + y_3 + y_4) - 3y_1 - 4x_2 - 5y_3 - 4.5y_4 - 180z_2$ subject to all $x_i \geqslant 0$, all $y_i \geqslant 0$, $z_1 = 1$ if any $x_i > 0$, $z_2 = 1$ if any $y_i > 0$, $z_1 = 0$ if all $x_i = 0$, $z_2 = 0$ if all $y_i = 0$, and $.20(x_1 + x_2 + x_3 + x_4) = x_3$, $.20(x_1 + x_2 + x_3 + x_4) = x_4$, $x_1 \geqslant .05 (x_1 + x_2 + x_3 + x_4)$, $y_1 = y_2 = y_3 = y_4$, $x_1 + y_1 \leqslant 1000$, $x_2 + y_2 \leqslant 900$, $x_3 + y_3 \leqslant 800$, and $x_4 + y_4 \leqslant 700$.

We will solve the two nontrivial equations and then write the program to maximize profit for the Saint Gervais Company. Rearranging the first two equations we get:

$$x_1 + x_2 + x_3 + x_4 = 5x_3$$
$$x_1 + x_2 + x_3 + x_4 = 5x_4$$

or

$$x_1 + x_2 - 4x_3 + x_4 = 0$$
$$x_1 + x_2 + x_3 - 4x_4 = 0$$

or

$$\begin{array}{rrrrr}
1 & 1 & -4 & 1 & 0 \\
1 & 1 & 1 & -4 & 0 \\
1 & 1 & -4 & 1 & 0 \\
0 & 0 & 5 & -5 & 0 \\
1 & 1 & -4 & 1 & 0 \\
0 & 0 & 1 & -1 & 0 \\
1 & 1 & 0 & -3 & 0 \\
0 & 0 & 1 & -1 & 0
\end{array}$$

or

$$x_1 + x_2 - 3x_4 = 0$$
$$x_3 - x_4 = 0$$

or for arbitrary x_2 and x_4

$$x_1 = 3x_4 - x_2$$

$$x_3 = x_4$$

The program follows:

```
4     REM SAINT GERVAIS PAINT CO.
5     REM BLENDING PROBLEM
6     REM WITH FIXED COSTS.
20    B=-999999
25    N=0
30    X=1
35    FOR I=1 to 50000
40    X4=INT(RND(X)*701)
45    X3=X4
50    X2=INT(RND(X)*901)
55    X1=3*X4-X2
58    IF X1 < 0 THEN 40
60    Y4=INT(RND(X)*(701-X4))
65    Y3=Y4
70    Y2=Y4
75    Y1=Y4
80    IF X1+Y1 >  1000 THEN 330
90    IF X2+Y2 >  900 THEN 330
100   IF X3+Y3 >  800 THEN 330
110   IF X1 < .05*(X1+X2+X3+X4) THEN 330
120   N=N+1
125   C1=6*X1+7*X2+8*X3+7.5*X4+300
126   C2=3*Y1+4*Y2+5*Y3+4.5*Y4+180
130   P1=12*(X1+X2+X3+X4)-C1
140   P2=9*(Y1+Y2+Y3+Y4)-C2
150   P=P1+P2
160   IF X1=0 THEN 190
170   IF Y4=0 THEN 210
180   GO TO 220
190   P=P+300
200   GO TO 170
210   P=P+180
220   IF P > B THEN 240
230   GO TO 330
240   A1=X1
250   A2=X2
260   A3=X3
270   A4=X4
280   A5=Y1
```

```
290  A6=Y2
300  A7=Y3
310  A8=Y4
320  B=P
330  NEXT I
340  PRINT A1,A2,A3,A4
350  PRINT A5,A6,A7,A8
360  PRINT B
370  STOP
380  END
```

```
964           899           621           621
0             0             0             0
15257.5

548           568           372           372
328           328           328           328
15206

966           897           621           621
0             0             0             0
15259.5
```

From these three runs of 50,000 answers we decide to try fixing $y_1 = y_2 = y_3 = y_4$ to zero and rerun. Four runs of 50,000 answers produce this printout:

```
1000   896   632   632

0      0     0     0

15552
```

It could be checked more, but this is quite probably the optimal solution. So 1000 liters of red pigment, 896 liters of A varnish, 632 liters of B thinner, and 632 liters of type D5 driers should be blended to make only Red 1 fireproof paint (Red 2 is not made). The resulting profit is $15,552.

However, Mr. Bordeaux, who is in charge of this computer project, is a little concerned that when the optimal solution is presented to his boss, the boss may say that both Red 1 and Red 2 paints should be made. Noticing the first printout yielded a $15,206 profit with $y_1 = y_2 = y_3 = y_4 = 328$ liters, Mr. Bordeaux reruns the program with the following line inserted:

```
60 Y4 = INT(RND(X)*51) + 300
```

This bounds the y_i's between 300 and 350. A few runs of this and more focusing yielded:

691	482	391	391
309	309	309	309
15425			

Therefore, $x_1 = 691$, $x_2 = 482$, $x_3 = 391$, and $x_4 = 391$ liters could go into Red 1 paint, and $y_1 = y_2 = y_3 = y_4 = 309$ liters could go into Red 2 paint to yield a profit of \$15,425. This is only \$127 below the optimal solution and might be a good alternative if the company wants to make both paints.

Another interesting idea would be to investigate demand as a function of price (develop price-quantity curves for Red 1 and Red 2) and rerun the program to see what the optimum would be. Also, if demand might be projected as far greater than the somewhat over 3000 liters of paint produced, then the company could investigate discounts for buying in quantity versus the cost of tying up the capital in an inventory of paint ingredients.

EXERCISES

8.1 The Newport Co. makes three types of filing cabinets. Type 1 sells for \$100, Type 2 for \$140, and Type 3 for \$190. The fixed advertising costs for this year are \$2400 for Type 1, \$2600 for Type 2, and \$3000 for Type 3. The unit assembly costs are \$15, \$17, and \$22, respectively, for the first 5000 units of each, and one and a half times that cost (due to overtime) for each unit over 5000.

All three filing cabinets use the same drawer assembly. However, Type 1 has one drawer (and hence requires one drawer assembly), Type 2 has two drawers, and Type 3 has three drawers. Newport buys 3 different sizes of sheet steel and bends them into the 3 different cabinet sizes. See the price lists below.

The bank will loan Newport \$1,000,000 at 10.75% interest for one year for this venture. Find the maximum profit production level given the price discounts and capital limitations.

Sheet steel for Cab. 1	Unit price (\$)	Sheet steel for Cab. 2	Unit price (\$)	Sheet steel for Cab. 3	Unit price (\$)
0 - 999	7.00	0 - 1199	8.50		
1000 - 5999	6.50	1200 - 3999	8.00	0 - 6999	10.00
6000 - 9999	6.25	4000 or over	7.50	7000 or over	9.00
10,000 or more	6.00				

Drawer assembly (one drawer) for any of the three filing cabinets	Unit price ($)
0 - 4999	20.00
5000 - 14,999	18.50
15,000 - 24,999	18.00
25,000 or more	17.50

8.2 The Brenner Manufacturing Group Ltd. is budgeting for production of its 4-meter outboard boat called the Canadian Northern.

A Canadian Northern boat consists of the aluminum boat which Brenner bends and molds from raw aluminum sheets, two wooden seats, an outboard, and a trim package (paint, decals, and accessories). The price-quantity lists for these four parts of each boat are listed below. The price-quantity curve for the boat is $P_1 = -.000009x_1 + 1690$. The company will borrow the money at 11.5% to finance purchases of the products needed to make the boat. In-house assembly costs are $100 per boat for the first 25,000 boats and $150 per boat for any boats over 25,000 units (due to overtime). Advertising costs will be fixed at $10,000 for the year. The company has such good credit that the bank will loan them the full amount they request. Also, management is concerned about price increases by their suppliers next year. Therefore, they want 10,000 Canadian Northern boats left in inventory at the end of the year. Find the maximum profit solution.

Sheet alum. for boat	Unit price ($)	Two wooden seats	Unit price ($)
0 - 4999	49.00	0 - 2999	10.00
5000 or more	40.00	3000 - 17,999	9.00
		18,000 - 29,999	8.50
		30,000 or more	8.00

Outboard	Unit price ($)	Trim Package	Unit price ($)
0 - 499	160	0 - 999	68.00
500 - 1999	140	1000 - 4999	62.00
2000 - 4999	135	5000 - 42,999	57.00
5000 - 14,999	130	43,000 or more	50.00
15,000 - 29,999	125		
30,000 or more	115		

8.3 The town of Eagle Mountain is closing its only fire station and plans to build two new ones to serve the community. Ten sites have been selected as possible locations for one of the two stations; each one is in a different section of the city. The ten sections of the city, their distances from each other (proposed site to center of the sections), and the number of fires in the past five years in each section are listed in the chart below. Determine the two sites the city should select for the stations in order to minimize the overall response distance to a fire anywhere in the city. (It is assumed that the station nearest the fire will respond.)

	Distance from site in section to middle of section J (in miles)									No. of fires in section in
Section	1	2	3	4	5	6	7	8	9	last five years
1										27
2	1.8									18
3	2.2	2.0								9
4	1.3	1.7	1.7							71
5	1.3	1.5	1.9	1.1						48
6	1.6	2.4	3.6	1.9	1.8					14
7	2.1	1.6	4.2	2.6	1.4	2.2				62
8	.8	.9	1.8	2.1	.8	1.9	1.1			59
9	.4	1.4	2.9	.9	1.9	1.3	1.1	1.7		43
10	2.3	2.0	2.2	.7	1.6	2.1	2.5	1.6	1.3	51

8.4 The U.S. Port inspector will host a conference for the Port Authority. Parties from the ten major U.S. ports will attend (listed below). The conference will last four days and the government will pay all expenses of the attendees. Therefore, the U.S. Port inspector wants to select one of these ten port cities as the site in such a way as to minimize the total expenses the government must pay. Use of the minimum cost solution will also keep him from being criticized for not awarding the conference to the nine cities that do not get the conference. All ten cities are lobbying for it. Find the minimum cost solution.

4-day conference cost per day ($)	Port Authority party	Port
70	7	New York
62	8	New Orleans
50	9	Houston
50	6	Baltimore
60	5	Philadelphia
55	3	Chicago
70	7	San Francisco
49	2	Norfolk
40	5	Duluth
40	4	Superior

Note: Although these are the ten largest U.S. ports, the costs are fictitious.

Round-Trip Transportation ($)

	N.Y.	N.O.	Houst.	Balt.	Phila.	Chi.	S.F.	Nor.	Dul.	Sup.
New York		300	310	70	49	170	400	130	205	200
New Orleans			105	300	300	290	280	250	240	240
Houston				280	300	250	300	290	270	280
Baltimore					50	190	400	40	200	210
Philadelphia						170	400	58	170	190
Chicago							270	159	100	90
San Francisco								410	275	300
Norfolk									250	260
Duluth										5
Superior										

8.5 The Bois Blanc Corporation is taking one containerized box out to the airport for shipment to Markleville. The plane only has room for the one box with dimensions of 2 meters by 2 meters by 5 meters. Therefore, they have a 20 cubic meter box to deal with. The weight constraint for the box is no more than 500 kilograms. The potential cargo with profit, size, and weight constraints are given below. Find the maximum profit solution.

Cargo	Maximum no. of units	Unit profit ($)	Unit size (cubic meters)	Unit weight (kilograms)
A	2	75	1.5	80
B	3	119	2.7	81
C	1	78	1.3	52
D	5	92	.9	41
E	2	162	2.8	39
F	4	101	1.5	62

8.6 Gladstone Limited has a blending problem. They make Grade A and Grade B fertilizer from three ingredients. The price-quantity curve for A is $P_A = -.00002 x_1 + 70$. The price-quantity curve for B is $P_B = -.00001x_2 + 55$. One bag of A consists of not more than 20% of Ingredient 1 and not less than 40% of Ingredient 3. One bag of B must contain at least 70% of Ingredients 1 and 2 combined and at least 25% of 1 and 2 individually.

Ingredient	Cost per bag ($)	Bag same size (as A & B bag)
1	10	\$9 if over 10,000 bags
2	12	\$11 if over 15,000 bags
3	13	\$10 if over 38,000 bags

The in-house blending costs are \$3 per bag. Find the maximum profit blending solution and the two prices Gladstone should charge to realize that profit.

8.7 The Copper Harbor Corporation of Australia has a financial planning problem. They can buy an aluminum distributorship for \$1.6 million. This is expected to earn 10% of the original investment (compounded—because aluminum is forecasted to boom) over the next five years. However, if they buy this distributorship they must buy a fleet of trucks for \$.4 million for deliveries. They can also buy a cannery for \$1.1 million. If they buy the cannery, management feels they must also buy a fishing fleet for \$.21 million. The cannery is expected to return 12% a year but not compounded.

They are considering purchase of a pulp mill for \$1.35 million. However, they can buy the pulp mill only if they buy a medium-sized forest for \$5.2 million. The pulp mill is expected to yield 18% a year compounded. In addition, the forest is loaded with bauxite which they could mine and cut their aluminum costs by an extra 5% a year (5% more profit per year) due to the fact that the aluminum smelters would give them a better price in return for a long-term bauxite supply (used to make aluminum).

Copper Harbor can also buy an iron mine for \$5 million that is expected to return 12% a year. If they buy the iron mine, they can buy a railroad for \$.5 million which they can use to send the iron ore to the docks and on the return journey can bring fish from their fishing fleet to the cannery at a drop in cost of 5% for the cannery.

The company will borrow the money for five years to finance one or more of these operations, and the loan will be due in five years. They can borrow \$2 million at 10%, or \$4 million at 12%. They can borrow \$6 million at 15%, or \$10 million at 17%. Any excess unused loan can be invested in bonds at 9.5%.

What investment strategy should they choose to maximize their return at the end of five years?

8.8 The following simple example will illustrate how a linear investment model is set up.

A textile company is planning to build a spinning mill. Two alternative (mutually exclusive) projects are available, I_1 and I_2. In addition, the building of a cloth mill is considered, for which the existence of a spinning mill is a prerequisite; there are three alternative projects, I_3, I_4, and I_5. Estimates of net present values Y_j and initial outlays (building costs) A_j, in units of \$1,000,000, are given in the following table:

$j=$	1	2	3	4	5
Y_j	6	4	3	5	6
A_j	16	12	8	12	16

At least 30% of the total initial outlay (total building cost) is required to be covered by the company's equity capital, \$8,400,000.

Translated into linear programming terms, the problem is the following: $Y = 6x_1 + 4x_2 + 3x_3 + 5x_4 + 6x_5 =$ maximum, subject to $x_1 + x_2 \leqslant 1$, $x_3 + x_4 + x_5 \leqslant 1$, $x_3 + x_4 + x_5 \leqslant x_1 + x_2$, $8.4 \geqslant 0.3\,(\Sigma A_j x_j)$, i.e., $16x_1 + 12x_2 + 8x_3 + 12x_4 + 16x_5 \leqslant 28$, and $x_j = 0$ or 1 $(j = 1, 2, \ldots, 5)$ or $x_j \leqslant 1, x_j \geqslant 0$ $(j = 1, 2, \ldots, 5)$ where (24) expresses the restriction that a spinning mill is a prerequisite of the cloth mill.*

*Exercise 8.8 is from p. 74 of *Linear Programming in Industry*, 4th ed, by Sven Dano. Reprinted here with permission of the publisher, Springer-Verlag, New York, 1974.

Suggested Reading

Budnick, Mojena, Vollmann. *Principles of Operations Research for Management.* Homewood, IL: Richard D. Irwin, 1977.

Dano, Sven. *Linear Programming in Industry*, 4th ed. New York: Springer-Verlag, 1974.

Gordon, G. and Pressman, I. *Quantitative Decision-Making for Business.* Englewood Cliffs, N.J.: Prentice-Hall, 1978.

Levin, R. and Lamone, R. *Linear Programming for Management Decisions.* Homewood, IL: Richard D. Irwin, 1969.

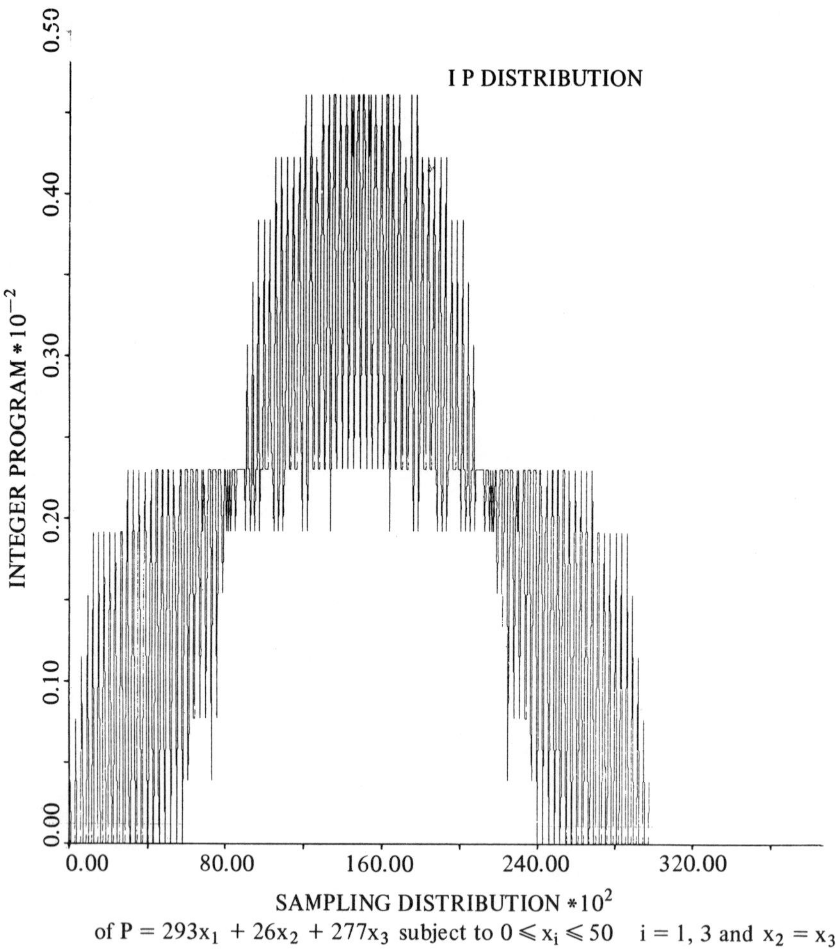

of $P = 293x_1 + 26x_2 + 277x_3$ subject to $0 \leqslant x_i \leqslant 50$ $i = 1, 3$ and $x_2 = x_3$

FIGURE 9.1 is another sampling distribution of all the answers of an optimization problem. There appear to be well defined trails to the optimal. We will exploit this with multi stage in chapter 12.

CHAPTER 9

Packaging and Shipping

The packaging industry has become very cost conscious with the increase in prices of most packaging materials. Along with this, shipping charges have been rising steadily with increased labor and fuel costs. With this in mind, designing packaging to minimize the cost of the particular container and/or to reduce its unit shipping cost is attractive in many instances.

Many factors influence the cost of packaging and shipping. Virtually all of them could be handled in a fairly massive packaging plant optimization model that would draw on many of the problems presented in other parts of the text.

Right now, let us look at a few package design and shipping examples.

EXAMPLE 9.1

The Telemark Packaging Company makes rectangular tin containers of inside size of one cubic meter (one million cubic centimeters or 1000 liters). Because of industrial requirements of its major buyer, Wesport Limited, one of the dimensions of the container must be 500.5 centimeters. The tin is .25 centimeters thick. One other side must be between 700 and 1700 centimeters in order to allow ease of loading. Design the rectangular container to fit these requirements and yet minimize the surface area in order to cut costs because the company's major problem is the cost of tin.

We want to minimize C = 500.5 x L x 2 + 500.5 x W x 2 + W x L x 2 subject to 700 $\leqslant L \leqslant$ 1700, and 500 x (L – .5) x (W –.5) = 1,000,000 cubic centimeters. The program follows:

```
5     B=999999
10    FOR L=700 TO 1700 STEP .1
15    W=1000000/(500*(L-.5)+.5
20    C=2*(500.5*L+500.5*W+W*L)
30    IF C < B THEN 50
40    GO TO 90
50    H1=500.5
60    L1=L
70    W1=W
80    B=C
90    NEXT L
100   PRINT H1, L1, W1
110   PRINT B
120   STOP
130   END
```

Therefore, the company should make their one cubic meter tin rectangular containers with height 500.5 cm, length 700 cm, and width 3.359 cm, which will yield a minimum surface area of 708,765 cm^2.

EXAMPLE 9.2

The inventory control department of the Rapid River Company has a warehouse that is 20 meters wide by 50 meters long by 3 meters high. They need to stack as many as possible of Product M which is in rectangular boxes of outside dimensions .5 meter by .4 meter by .3 meter. How should they stack Product M boxes in order to get the most in the warehouse?

Assuming we use one of the six basic stacking formations (depending on which side the box is laid down), we can just do the divisions in this case to check:

	Warehouse			
	Width	Length	Height	
	20	50	3	
.5	40	100	6	Divisions
.4	50	125	7	rounded
.3	66	166	10	down

So we have:

$$40 \times 125 \times 10 = 50{,}000$$

$$40 \times 166 \times 7 = 46{,}480$$

$$100 \times 50 \times 10 = 50{,}000$$

$$100 \times 7 \times 66 = 46{,}200$$

$$6 \times 125 \times 66 = 49{,}500$$

$$6 \times 166 \times 50 = 49{,}800$$

Either .5 to width, .4 to length, and .3 to height, or .5 to length, .4 to width, and .3 to height will allow a maximum of 50,000 units to be stacked in the warehouse.

This problem gets much more interesting (and requires a computer program) if two or three products are stacked in the same warehouse or boxcar with each having its own stacking formation with the objective again to maximize use given certain constraints. (See exercise 9.2.)

EXAMPLE 9.3

The Skowhegan Company will be making regular shipments of its product by containerized boxes of dimensions (inside) 2 by 3 by 8 meters. The company's product is put in a box of volume 10,000 cubic centimeters. Then the boxes are stacked in one of the six basic stacking formations in the containerized box.

The company will make the 10,000 cubic centimeter boxes. Since the shipping charges are substantial, it is desired to find the dimensions for the 10,000 cubic centimeter boxes that will allow the company loaders to put the maximum number in each container (using one of the basic stacking formations), hence reducing shipping costs.

The height and/or width of the box cannot exceed 50 centimeters. Each box dimension must be at least 4 centimeters. Also, the walls of the boxes are 1/2 centimeter on all six sides.

The program below looks at all combinations of heights and widths by one centimeter increments (more accuracy could be obtained from a finer mesh) and all stacking formations for each. It saves and prints the optimum dimensions and stacking formation.

```
1     REM A PROGRAM TO FIND THE
2     REM DIMENSIONS FOR A RECTANGULAR
3     REM BOX OF VOLUME 10,000 CUBIC
4     REM CENTIMETERS SUCH THAT THE
```

```
5    REM MAXIMUM NUMBER OF THESE BOXES,
6    REM THAT CAN BE STORED IN A
7    REM CONTAINER 2 BY 3 BY 8 METERS.
8    REM WILL BE ACHIEVED.
9    REM SIDE CONDITIONS: THE HEIGHT
10   REM AND/OR WIDTH OF THE BOX
11   REM CANNOT EXCEED 50 CENTIMETERS
12   REM ALSO THE WALLS OF THE INDIVIDUAL
13   REM BOXES ARE ONE HALF CENTIMETER
14   REM ON ALL SIX SIDES.
15   REM UNIFORM STACKING OF THE BOXES,
16   REM WITH ONE OF THE SIX BASIC
17   REM STACKING CONFIGURATIONS, IN
18   REM THE CONTAINER IS ASSUMED.
19   REM EACH DIMENSION MUST BE AT
20   REM LEAST 4 CENTIMETERS.
30   B=-999999
40   FOR I=4 TO 50
50   H=I
60   FOR J=4 TO 50
70   W=J
80   L=10000./(H*W)
90   REM H INTO 200 W INTO 300
100  N2=INT(200/H+1))
110  N3=INT(300/(W+1))
120  N8=INT(800/(L+1))
130  M1=N2*N3*N8
140  IF M1 > B THEN 160
150  GO TO 220
160  A2=N2
170  H1=H
180  A3=N3
190  W1=W
200  A8=N8
210  L1=L
215  B=M1
220  REM W INTO 200 H INTO 300
230  N2=INT(200/(W+1))
240  N3=INT(300/(H+1))
250  N8=INT(800/(L+1))
260  M2=N2*N3*N8
270  IF M2 > B THEN 290
280  GO TO 360
290  A2=N2
300  H1=H
310  A3=N3
320  W1=W
330  A8=N8
340  L1=L
```

```
350  B=M2
360  REM L INTO 200 W INTO 300
370  N2=INT(200/(L+1))
380  N3=INT(300/(W+1))
390  N8=INT(800/(H+1))
400  M3=N2*N3*N8
410  IF M3 > B THEN 430
420  GO TO 500
430  A2=N2
440  H1=H
450  A3=N3
460  W1=W
470  A8=N8
480  L1=1
490  B=M3
500  REM W INTO 200 L INTO 300
510  N2=INT(200/(W+1))
520  N3=INT(300/(L+1))
530  N8=INT(800/(H+1))
540  M4=N2*N3*N8
550  IF M4 > B THEN 570
560  GO TO 640
570  A2=N2
580  H1=H
590  A3=N3
600  W1=W
610  A8=N8
620  L1=L
630  B=M4
640  REM H INTO 200 L INTO 300
650  N2=INT(200/(H+1))
660  N3=INT(300/(L+1))
670  N8=INT(800/(W+1))
680  M5=N2*N3*N8
690  IF M5 > B THEN 710
700  GO TO 780
710  A2=N2
720  H1=H
730  A3=N3
740  W1=W
750  A8=N8
760  L1=L
770  B=M5
780  REM L INTO 200 H INTO 300
790  N2=INT(200/(L+1))
800  N3=INT(300/(H+1))
810  N8=INT(800/(W+1))
820  M6=N2*N3*N8
830  IF M6 > B THEN 850
```

```
840  GO TO 920
850  A2=N2
860  H1=H
870  A3=N3
880  W1=W
890  A8=N8
900  L1=L
910  B=M6
920  NEXT J
930  NEXT I
940  PRINT A1,H1, A3,W1,A8,L1
950  PRINT B
955  REM DO THE DIVISIONS TO FIND OUT
957  REM WHICH DIMENSION IS WHICH
960  STOP
970  END

8          14         20          24          26
29.7619
4160
```

IH=14 centimeters for height, and 14+1 into 300=20=IN300. So the height of the box is 14 and lay that dimension down across the 3-meter direction. IW=24 centimeters for width, and 24+1 into 200=8=IN200. So the width of the box is 24 and lay that dimension down across the 2-meter direction. This leaves RL=29.762 centimeters for length to be stacked in the 8-meter direction. This gives 8 x 20 x 26 = 4160 boxes per container.

EXAMPLE 9.4

The Grand Junction Company has been shipping 40 units of A and 40 units of B in one 3.2 meter wide, 3.15 meter high, and 12.2 meter long boxcar once a day from their West Coast plant to their Roanoke, Virginia warehouse for overseas shipping. They want to keep using these boxcars to send about the same amount of A and B each day. However, they wonder if they could get more units in the boxcar.

One unit of A is in a .5 x 1 x 2 meter rectangular box (outside dimensions). A unit of B is in a .4 x .8 x 3 meter rectangular box (outside dimensions). Therefore, the company decides to stack the units of B with the 3-meter side standing up (3 into 3.15 equals one plus), the .4-meter side lying widthwise (.4 into 3.2 equals eight exactly), and the .8 side lying down lengthwise. They will lay 7 down lengthwise. Therefore, 1 x 8 x 7

= 56 units of B will be put in the boxcar. This means the units of A will have to fit in a 3.2 x 3.15 x (12.2 -5.6) = 6.6 meter rectangular space.

	3.2	3.15	6.6
.5 into	6	6	13
1 into	3	3	6
2 into	1	1	3

Thus

6 x 3 x 3 = 54

6 x 1 x 6 = 36

3 x 6 x 3 = 54

3 x 1 x 13 = 39

1 x 3 x 13 = 39

1 x 6 x 6 = 36

Therefore, the .5 side goes widthwise, the 1 side goes up, and the 2 side goes lengthwise, yielding 54 A units stacked in the boxcar. Of course, there could be other stacking formations, but this one will yield 56 + 54 = 110 units per car instead of the old inefficient way of stacking.

EXERCISES

9.1 In Example 9.4 (Grand Junction's A and B boxcar problem) we were able to get 110 units in the car, each with a volume of 1 meter. But from the dimensions of the boxcar, we get that there are 3.2 x 3.15 x 12.2 = 122.976 cubic meters in the boxcar. Therefore, we could perhaps get in 12 more units if we redesigned the 1 cubic meter A and B boxes. Do this.
Hint: This could be difficult. Perhaps pin down a few dimensions like 3.15 meters high for boxes A and B and then go to work on the rest of them.

9.2 Grand junction is also designing a 2 cubic meter container (rectangular) to fit in a 2.8 x 3.15 x 18.3 meter (extra long size) boxcar. Design the rectangular container and stacking configuration to allow the maximum number to be placed in a boxcar.

9.3 The Bavarian Packaging Company has a 1.5 cubic meter square box (outside dimensions) which it sells to many customers. These customers ship their products in this box in railroad boxcars. However, the railroad has three standard-size boxcars that they use:

	Boxcar sizes (in meters)		
	Regular	Long	Extra Long
Width	2.8	2.8	2.8
Height	3.15	3.15	3.15
Length	12.2	15.25	18.3

Find the optimal stacking formation for each size railroad car so that Bavarian can include these instructions with a shipment of boxes to their customers and can also advertise and promote their efficiency.

9.4 The Bolzano Company is shipping an industrial chemical in a 100-liter cylindrical can. (Inside and outside dimensions are virtually the same–metal is very thin.) But the cylindrical cans do not stack well. Redesign the container to minimize the surface area and make it rectangular. (*Hint:* 100 liters is 100,000 cubic centimeters.)

9.5 Design a rectangular metal drum of 2.5 cubic meters (outside dimensions) so that a maximum number of these can be stacked in the extra long railroad cars from exercise 9.3.

9.6 The Japanese Can Company makes a 1000 cubic centimeter (outside dimensions) rectangular container. The sides are made of pressed cardboard and cost 1 cent per cubic centimeter. The lip (top) costs only .6 cents per cubic centimeter, but the base costs 1.4 cents per cubic centimeter. Also, length and width must be between 5 and 15 centimeters, and the height must be between 8 and 12 centimeters. Design the container to minimize the cost.

9.7 Once in a while the 2 cubic meter container from exercise 9.2 has to be shipped in a rectangular semi-truck (with cargo space dimensions of 3 meters by 5 meters by 10 meters). Find the optimal stacking pattern for the 2 cubic meter container.

9.8 Rewrite the Skowhegan program from Example 9.3 using loops and subscripted variables to condense it to about one-third of its length.
Hint: In BASIC the statement 5 DIM X(100), for example, would declare

X(1), X(2), X(3), . . . , X(100)

one hundred subscripted variables for use later in the program. Then, say

```
10 FOR I = 1 TO 100
20 READ X(I)
30 NEXT I
```

would assign the next 100 data values (from a data statement) to X(1), X(2), . . . , X(100) in order. Another example, 20 DIM Y(8), gives the variables Y(1), Y(2), Y(3), Y(4), Y(5), Y(6), Y(7), Y(8) to use.

Suggested Reading

Ceder, J. and Outcalt, D. *A Short Course in Calculus.* New York: Worth Publishers, 1968.

Graworg, Fielitz and Robinson, Tabor. *Mathematics: A Foundation for Decisions.* Reading, Mass.: Addison-Wesley, 1976.

Laufer, Arthur. *Operations Management.* Cincinnati: Southwestern Co., 1975.

Miller, Ronald. *Modern Mathematical Methods for Economics and Business.* New York: Holt, Rinehart and Winston, 1972.

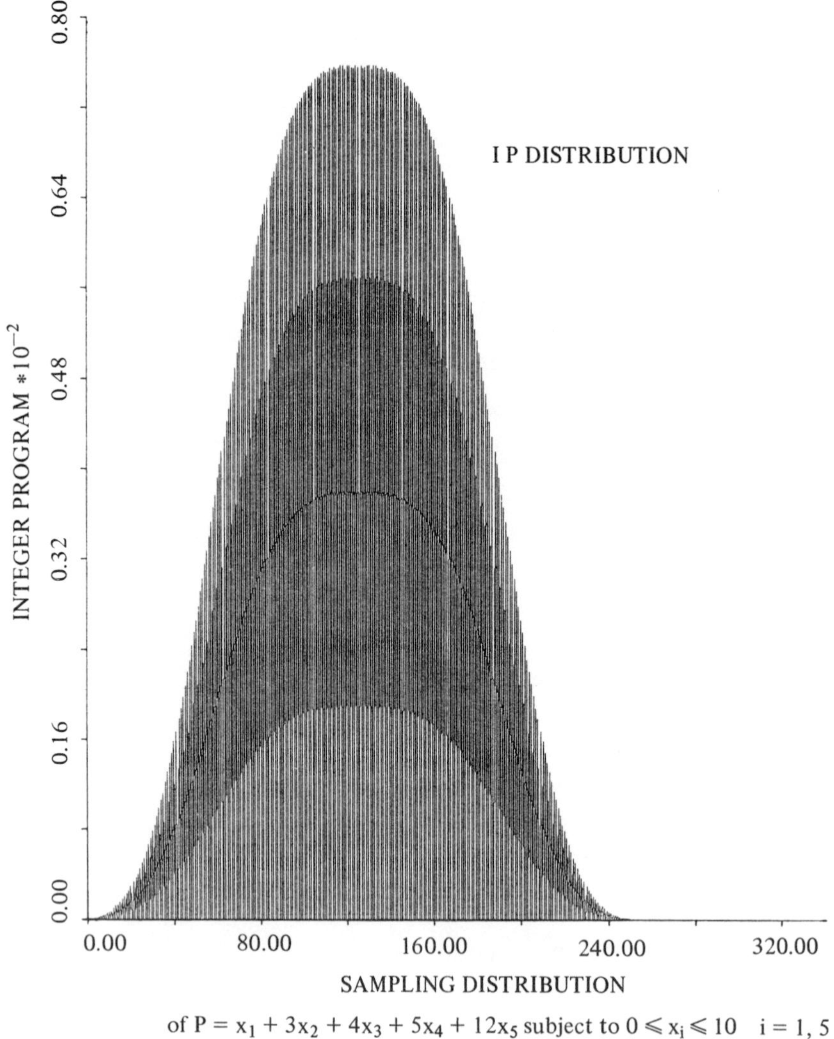

of $P = x_1 + 3x_2 + 4x_3 + 5x_4 + 12x_5$ subject to $0 \leqslant x_i \leqslant 10 \quad i = 1, 5$

FIGURE 10.1

CHAPTER 10

Statistical Justification of Computer Optimization

It is the author's contention that virtually any optimization problem of one hundred variables or less can be solved with today's computers. The most sophisticated and efficient programs to do this are the multistage Monte Carlo integer programs which are presented and discussed in chapter 12. However, even without these, repeated simulations and focus searches and/or searching the feasible solution space by increments of 10, 100, or 1000, etc., should produce the true optimum. The major advantage of the Monte Carlo optimization approach is that it frees the user to write accurate nonlinear models and not be held back by linearity, continuity, and/or differentiality assumptions.

However, even with all of these advantages, let us look at Monte Carlo optimization statistically. Throughout the text there are sampling distributions of selected integer programs; for example, figures 10.1 and 10.2 are graphs of all the feasible solutions of two integer programming problems. The optimum (maximum) answer is at the extreme right of the distribution graph. The most frequently occurring answer is represented by the highest peak. These distributions are generated by treating the objective (say, profit) function as a statistic and evaluating the function for each feasible solution, plotting the frequency of occurrence of each answer. If Monte Carlo optimization is to work well, the important thing is for the optimal solution not to be too far isolated from the rest of the feasible solutions. After a few Monte Carlo runs one can then

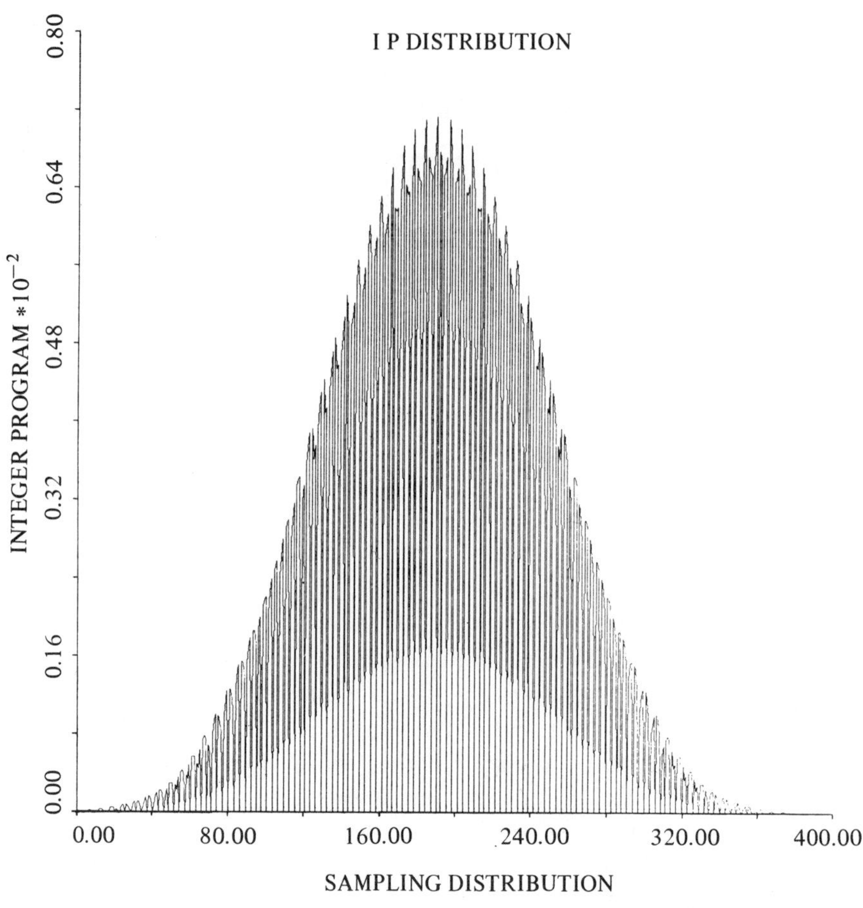

of $P = 6x_1 + 7x_2 + 12x_3 + 7x_4 + 6x_5$ subject to $0 \leqslant x_i \leqslant 10 \quad i = 1, 5$

FIGURES 10.1 and 10.2. A program designed to travel across the distribution in question until it finds the optimal solution works as follows: A sample of between 100 and 5000 is drawn (using the random number generator). The maximum of this sample is found. This "optimal so far" will be in the upper .01 of the distribution, for example. Then another sample of between 100 and 5000 is drawn in a narrower region about this "optimum so far" (each time recentering the region about any better "optimums so far" produced). This new answer would be in, say, the upper .0001 of the distriubtion. Then another and another sample, etc. are taken (with ever narrowing bounds and appropriate centering with each new "optimum so far") and the program funnels into the right answer. On a problem with 1×10^{40} answers or less, this process takes only a few minutes or several seconds on a computer. (Figure 10.2 reprinted by permission of the National Bureau of Standards, Special Publication 503, p. 365.)

focus on the optimum solution area and find the true optimum solution. What would hurt the Monte Carlo approach the most would be a distribution (in an optimization problem) that had one isolated optimal solution that was far away from the rest of the solutions as in Figure 10.3.

After studying thousands of these distributions, the author believes that this will virtually never happen in a practical optimization problem. Notice that the distributions in the text are well connected, no isloated optimums. For the sake of illustration, consider the standard distribution of an optimization problem in Figure 10.4. The area under the curve or

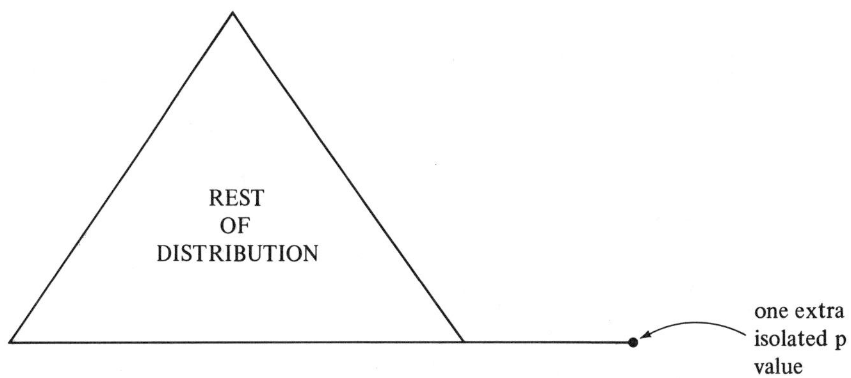

FIGURE 10.3

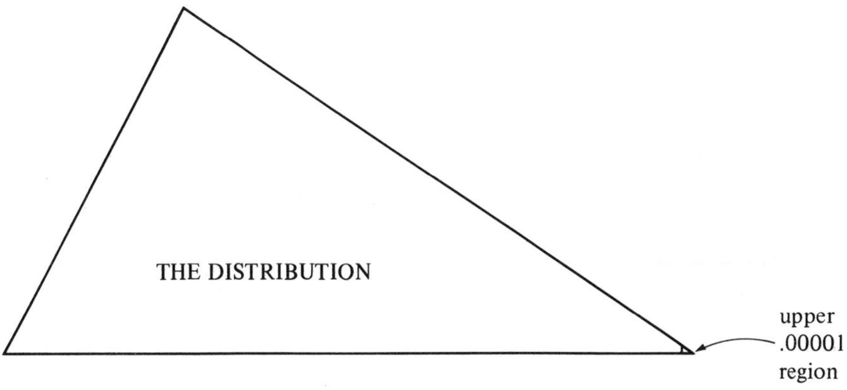

FIGURE 10.4

the sum of all the probabilities of each histogram class is one. Therefore, the question could be asked, What is the probability that at least one solution in a Monte Carlo optimization run of 1,500,000 solutions will be in the upper .00001 region of the distribution and hence nearly optimum? This probability is equal to one minus the probability that none of the 1,500,000 solutions lands in the upper .00001 region or $1 - .99999^{1,500,000} = 1 - .0000003058 = .9999996942$. Therefore, the odds are overwhelming that a nearly optimal solution will be found.

EXAMPLE 10.1

Consider the problem to maximize $P = 10x_1{}^2 + 30x_2 + 50x_3 + 70x_4 + 90x_5$ subject to $0 \leqslant x_i \leqslant 200$ for $i = 1, 2, \ldots, 5$. Now the optimal solution is $x_1 = x_2 = x_3 = x_4 = x_5 = 200$ with $P = 448{,}000$. But let us write a Monte Carlo program to try to randomly look for the answer:

```
5      REM MONTE CARLO INTEGER PROGRAM
7      REM NONLINEAR PROBLEM
10     B=-999999
20     X=1
30     FOR I=1 TO 100000
40     X1=INT(RND(X)*201)
50     X2=INT(RND(X)*201)
60     X3=INT(RND(X)*201)
70     X4=INT(RND(X)*201)
80     X5=INT(RND(X)*201)
90     P=10*X1**2+30*X2+50*X3+70*X4+90*X5
100    IF P  >  B THEN 120
110    GO TO 180
120    A1=X1
130    A2=X2
150    A3=X3
155    A4=X4
160    A5=X5
170    B=P
180    NEXT I
190    PRINT A1,A2,A3,A4,A5
200    PRINT B
210    STOP
220    END

200           162           176           197           198
445270

200           133           199           198           193
445170
```

```
200         133         199         197         196
445370

200         199         196         191         184
445700

200         192         153         191         173
442350

200         192         154         198         199
445230

200         200         198         194         182
445860

200         200         198         191         164
444030

200         192         153         196         197
44860

200         193         156         200         200
445590
```

After these runs it was decided to bound x_2 by 130 and 200, x_3 by 150 and 200, x_4 by 190 and 200, and x_5 by 180 and 200. Due to an oversight, x_1 was not focused, but it did not matter; the optimal solution was still found (see printout below):

```
5     REM MONTE CARLO INTEGER PROGRAM
7     REM NONLINEAR PROBLEM
8     REM FOCUS SEARCH
10    B=-999999
20    X=1
30    FOR I=1 TO 100000
40    X1=INT(RND(X)*201)
50    X2=INT(RND(X)*71)+130
60    X3=INT(RND(X)*51)+150
70    X4=INT(RND(X)*11)+190
80    X5=INT(RND(X)*21)+180
90    P=10*X1**2+30*X2+50*X3+70*X4+90*X5
100   IF P > B THEN 120
110   GO TO 180
120   A1=X1
130   A2=X2
150   A3=X3
155   A4=X4
```

```
160  A5=X5
170  B=P
180  NEXT I
190  PRINT A1,A2,A3,A4,A5
200  PRINT B
210  STOP
220  END

200        200        200        200        199
447910

200        200        199        200        199
447860

200        200        200        200        199
447910

200        200        199        200        198
447770

200        200        199        200        198
447770

200        200        199        200        199
447860

200        200        200        200        199
447910

200        199        196        197        198
447380

200        200        199        200        198
447770

200        200        200        200        200
448000
```

There were 201 x 201 x 201 x 201 x 201 = 328,080,401,000 solutions and yet the simulation produced the optimum quickly.

EXAMPLE 10.2

Now let us consider another approach. Maximize $P = 2x_1 + x_2{}^2 + 6x_3 + 5x_4 + 9x_5$ subject to $0 \leqslant x_i \leqslant 35$ for all i's, $x_1 + 7x_2 + 3x_3 + 2x_4 + x_5 \leqslant 360$, $x_1 + 2x_2 + 3x_3 + 4x_4 + 5x_5 \leqslant 400$, and $5x_1 + 4x_2 + 3x_3 + 2x_4 + x_5 \leqslant 400$.

There are $36^5 = 60{,}466{,}176$ potential answers (many of which will be thrown out for not satisfying the constraints). Therefore, we decide to search for the maximum answer by increments of 3 (0, 3, 6, 9, 12, . . . , 33) which now leaves only $12^5 = 28{,}832$ answers to look at. The program below does this:

```
10    B=-999999
15    N=0
20    FOR X1=0 TO 35 STEP 3
30    FOR X2=0 TO 35 STEP 3
40    FOR X3=0 TO 35 STEP 3
50    FOR X4=0 TO 35 STEP 3
60    FOR X5=0 TO 35 STEP 3
70    IF X1+7*X2+3*X3+2*X4+X5 > 360 THEN 190
80    IF X1+2*X2+3*X3+4*X4+5*X5 > 400 THEN 190
90    IF 5*X1+4*X2+3*X3+2*X4+X5 > 400 THEN 190
100   P=2*X1+X2**2+6*X3+5*X4+9*X5
105   N=N+1
110   IF P > B THEN 130
120   GO TO 190
130   A1=X1
140   A2=X2
150   A3=X3
160   A4=X4
170   A5=X5
180   B=P
190   NEXT X5
200   NEXT X4
210   NEXT X3
220   NEXT X2
230   NEXT X1
235   PRINT N
240   PRINT A1,A2,A3,A4,A5
250   PRINT B
260   STOP
270   END

232229
3            33           9            33           33
1611
```

After seeing this printout, we decide to focus and search all points in the following region: $0 \leqslant x_1 \leqslant 35$, $30 \leqslant x_2 \leqslant 35$, $0 \leqslant x_3 \leqslant 35$, $30 \leqslant x_4 \leqslant 35$, and $30 \leqslant x_5 \leqslant 35$. The program on the following page checks these 279,936 points (lines 30, 50, and 60 were changed).

```
5     REM FOCUS SEARCH FOR TRUE OPTIMUM
10    B=-999999
15    N=0
20    FOR X1=0 TO 35
30    FOR X2=30 TO 35
40    FOR X3=0 TO 35
50    FOR X4=30 TO 35
60    FOR X5=30 TO 35
70    IF X1+7*X2+3*X3+2*X4+X5 > 360 THEN 190
80    IF X1+2*X2+3*X3+4*X4+5*X5 > 400 THEN 190
90    IF 5*X1+4*X2+3*X3+2*X4+X5 > 400 THEN 190
100   P=2*X1+X2**2+6*X3+5*X4+9*X5
105   N=N+1
110   IF P > B THEN 130
120   GO TO 190
130   A1=X1
140   A2=X2
150   A3=X3
160   A4=X4
170   A5=X5
180   B=P
190   NEXT X5
200   NEXT X4
210   NEXT X3
220   NEXT X2
230   NEXT X1
235   PRINT N
240   PRINT A1,A2,A3,A4,A5
250   PRINT B
260   STOP
270   END

48023
1          35          3          35          35
1735
```

Therefore, the true maximum solution to this nonlinear integer programming problem—$x_1 = 1, x_2 = 35, x_3 = 3, x_4 = 35, x_5 = 35$ with $P = 1735$—was found with a minimum of effort.

This or the Monte Carlo approach allows us to solve problems of any type of ten variables or less easily and ten to twenty variables with a little effort.

EXAMPLE 10.3

Consider maximize $P = (EXP(.004x) + 16EXP(.008y) + z^{.25})/(x + y + 10z)$ subject to $x \geqslant 1, y \geqslant 1, z \geqslant 1, x + y + z \leqslant 200, 3x + 2y + z \leqslant 600,$

and $x + 2y + 3z \leqslant 600$. The program is below (note that lines 20 through 60 take care of the $x + y + z \leqslant 200$ constraint, among other things):

```
5     N=0
10    B=-999999
20    FOR X=1 TO 200
30    K=200-X
40    FOR Y=1 TO K
50    L=200-X-Y
60    FOR Z=1 TO L
70    IF 3*X+2*Y+Z > 600 THEN 180
80    IF X+2*Y+3*Z > 600 THEN 180
90    P1=EXP(.004*X)+16*EXP(.008*Y)+Z**.25
100   P2=X+Y+10*Z
110   P=P1/P2
115   N=N+1
120   IF P > B THEN 140
130   GO TO 180
140   B=P
150   A1=X
160   A2=Y
170   A3=Z
180   NEXT Z
190   NEXT Y
200   NEXT X
210   PRINT A1,A2,A3
220   PRINT B
230   PRINT N
240   STOP
250   END

1              1              1
1.51104
1313400
```

Therefore, the 1,313,400 feasible solutions were checked and produced an optimal solution of $x = 1$, $y = 1$, $z = 1$ with $P = 1.51104$. A focus search by increments of, say, .01 could produce a more accurate maximum (accurate to two decimal places) if necessary.

EXAMPLE 10.4

Minimize $C = (17x + x^3 + 12x^4)/(4x^2 + 42x^3 + .2x^5)$ subject to $1 \leqslant x \leqslant 50$. Find the minimum to the nearest two decimal places. The program is given on the following page.

```
5     REM ONE VARIABLE
6     REM MINIMIZATION
7     REM BY INCREMENTS OF
8     REM ONE ONE HUNDREDTH
10    B=999999
20    FOR X=1 TO 50 STEP .01
21    C1=17*X+X**3+12*X**4
22    C2=4*X**2+42*X**3+.2*X**5
30    C=C1/C2
40    IF C < B THEN 60
50    GO TO 80
60    A1=X
70    B=C
80    NEXT X
90    PRINT A1,B
100   STOP
110   END

1.38000          .584961
```

Therefore, $x = 1.38$ yields the minimum C value of .584961.

EXAMPLE 10.5

Consider the dynamic programming problem to maximize $P = 7x_1(3 - x_1) + 4x_2(4 - x_2)$ subject to $0 \leqslant x_1 \leqslant 30$ and $0 \leqslant x_2 \leqslant 15$. The program is:

```
5     REM TWO VARIABLE DYNAMIC
6     REM PROGRAMMING MODEL
7     REM WE WILL MAXIMIZE
8     REM WITH OUR TECHNIQUES
10    B=-999999
20    FOR X1=0 TO 30
30    FOR X2=0 TO 15
40    P=7*X1*(3-X1)+4*X2*(4-X2)
50    IF P > B THEN 70
60    GO TO 100
70    A1=X1
80    A2=X2
90    B=P
100   NEXT X2
110   NEXT X1
120   PRINT A1,A2
```

```
130  PRINT B
140  STOP
150  END

1              2
30
```

Therefore, $x_1 = 1, x_2 = 2$ with $P = 30$.

EXERCISES

10.1 What is the probability that a random sample of 1,000,000 feasible solutions will produce at least one answer in the upper .00001 region of the sampling distribution of all the answers?

10.2 Write and run repeatedly a Monte Carlo simulation to optimize $P = 100x_1 + 50x_2 + 60x_3 + 3x_4$ subject to $0 \leqslant x_i \leqslant 1000$. Then focus the search for the answer. The answer is obviously $P(1000, 1000, 1000, 1000) = 213{,}000$.

10.3 Write and run repeatedly a Monte Carlo simulation to optimize $P = 10x_1 + 90x_2 + 16x_3$ subject to $0 \leqslant x_1 \leqslant 100$, $0 \leqslant x_2 \leqslant 80$, and $20 \leqslant x_3 \leqslant 70$. The solution is, of course, $P = (100, 80, 70) = 9320$.

10.4 The price quantity curve for Product 1 is $P_1 = -.00001x_1 + 50$ with a unit cost of \$4 to produce. The price-quantity curve for Product 2 is $P_2 = -.00002x_2 + 100$ with a unit cost of \$3.50 to produce. However, plant capacity limits production in such a way that $2x_1 + 3x_2 \leqslant 10{,}000{,}000$. Also, $3x_1 + 2x_2 \leqslant 9{,}000{,}000$ from labeling and canning expenditures. Maximize the profit by finding the optimal combination of Product 1 and Product 2 to produce.

Suggested Reading

Greenberg, Michael. *Applied Linear Programming for the Socioeconomic and Environmental Sciences.* New York: Academic Press, 1978.

Larson, Harold. *Introduction to Probability Theory and Statistical Inference.* New York: John Wiley & Sons, 1969.

Simmonard, M. *Linear Programming.* Englewood Cliffs, N.J.: Prentice-Hall, 1966.

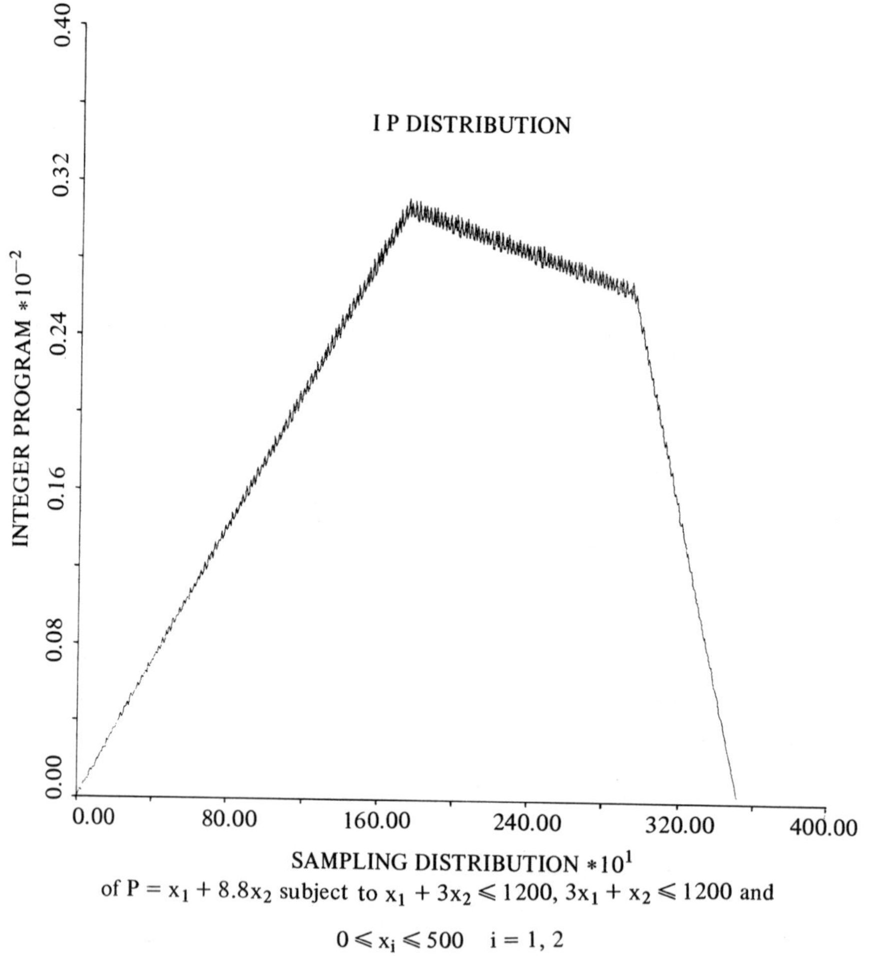

FIGURE 11.1. Even with a problem with constraints, because the answers are all connected, it is easy for the program to "slide across" the distribution and funnel into the true optimal solution.

CHAPTER 11

Linear Programming and Transportation Problems

Throughout the text we have been dealing with several types of optimization problems. One of the advantages of the search all points and/or simulation techniques is that the user does not have to classify the problem into a particular group and then master a new technique to solve it.

However, linear programming is an area of such sufficient importance (particularly because the simplex solution technique makes possible the computer solution of large-scale problems) that we shall discuss it here. A linear programming problem consists of a function of two or more variables to be maximized or minimized subject to a series of constraints. However, the function and all of the constraints must be linear (no squared or cubed terms or exponentials, etc.). An example would be maximize $P = 7x + 2y + 4z$ subject to $x \geqslant 0, y \geqslant 0, z \geqslant 0, 2x + y + 1.5z \leqslant 100$, and $3x + y + 2z \leqslant 130$. Additionally, fractional answers are acceptable, for example an optimal solution such as $x = 2/3, y = 16.5$, and $z = 48.1$.

Obviously, these restrictions can be limiting at times. If the $P = 7x + 2y + 4z$ above was a profit function, the $7x$ term would imply that the company would earn \$7 profit per unit of x produced. Realistically, it may be true that they will average \$7 per unit, but if they produced just one unit, would they earn \$7? Would they recover their fixed costs, design, planning, marketing, and transportation costs with just one unit produced?

Even with these limitations in some cases, linear programming has been and will continue to be an area of importance to business and government. Probably the major reason for its success is that the simplex solution technique (for solving linear programming problems) works on two- or two thousand-variable problems. Therefore, computer programs can be written to carry out these calculations quickly. This makes large-scale planning possible.

But in the rush to use the simplex solution technique, the user must be sure not to linearize a nonlinear optimization model if it would make the problem too inaccurate. This would produce the exact answer to the wrong question. At that point, it would be better to leave the model nonlinear and accurate and do a Monte Carlo simulation to find the optimum.

EXAMPLE 11.1

Now let us consider a linear programming problem. Maximize $P = 15x + 10y$ subject to $x \geqslant 0$, $y \geqslant 0$, $x + 2y \leqslant 25$, and $2x + y \leqslant 25$.

Notice that any x, y pair that is to be considered for the optimal solution must satisfy the four inequality constraints simultaneously. Let us graph these four inequalities to identify the feasible solution space; see Figure 11.2. The solution of $x + 2y = 25$ and $2x + y = 25$ will give the upper-right corner point of the feasible solution space. Thus

$$\begin{matrix} 1 & 2 & 25 \\ 2 & 1 & 25 \end{matrix}$$

$$\begin{matrix} 1 & 2 & 25 \\ 0 & -3 & -25 \end{matrix}$$

$$\begin{matrix} 1 & 2 & 25 \\ 0 & 1 & 25/3 \end{matrix}$$

$$\begin{matrix} 1 & 0 & 25/3 \\ 0 & 1 & 25/3 \end{matrix}$$

Therefore, $x = 25/3$, $y = 25/3$ is the upper-right point.

The fundamental theorem of linear programming states that the optimum solution (either maximum or minimum) occurs at a "corner" point of the feasible solution space. So in this case P is maximized at either:

(x, y)	P
(0,0)	0

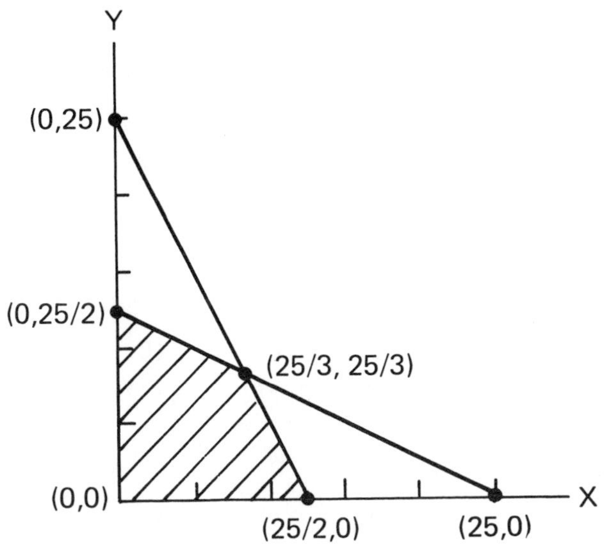

FIGURE 11.2

(25/2,0)	375/2 = 187.5
(25/3,25/3)	625/3 = 208.33
(0,25/2)	250/2 = 125

Therefore, the maximum is that $x = 25/3$, $y = 25/3$.

The simplex method of solution to a linear programming problem systematically finds the optimal solution (without graphing, etc.) by starting at one corner point and moving to another and another, and so on, always moving toward the optimum solution until it finds it. (A "corner" point in m dimensional space is always a solution of part of the related equalities that come from the constraints.)

EXAMPLE 11.2

With three or more variables the graphical approach becomes impractical, so the simplex technique is used. As a demonstration let us maximize $P = 4x_1 + 5x_2$ subject to $x_1 \geqslant 0$, $x_2 \geqslant 0$, $x_1 + x_2 \leqslant 10$, and $x_1 + 2x_2 \leqslant 15$ using the simplex technique.

Table 1

	x_1	x_2	x_3	x_4	constant		coefficient
P_i's	4	5	0	0			
	1	1	1	0	10	x_3	0
	1	(2)	0	1	15	x_4	0
Row 1	0	0	0	0	0		
P_i's-Row 1	4	5	0	0			

Table 2

x_1	x_2	x_3	x_4	constant
1	1	1	0	10
1/2	1	0	1/2	15/2

Table 3

	x_1	x_2	x_3	x_4	constant		coefficient
P_i's	4	5	0	0			
	(1/2)	0	1	-1/2	5/2	x_3	0
	1/2	1	0	1/2	15/2	x_2	5
Row 1	5/2	5	0	5/2	75/2		
P_i's-Row 1	3/2	0	0	-5/2			

Table 4

x_1	x_2	x_3	x_4	constant
1	0	2	-1	5
1/2	1	0	1/2	15/2

Table 5

	x_1	x_2	x_3	x_4	constant		coefficient
P_i's	4	5	0	0			
	1	0	2	-1	5	x_1	4
	0	1	-1	1	5	x_2	5
Row 1	4	5	3	1	45		
P_i's-Row 1	0	0	-3	-1	so done		

The answer is $x_1 = 5, x_2 = 5$ with a corresponding $P = 45$.

Explanation

In Table 1 the P_i's row consists of the coefficients of the profit function plus two extra 0 coefficients, one above each artificial variable x_3

and x_4. There are two artificial variables (in this case x_3 and x_4) because there are two nontrivial constraints: $x_1 + x_2 \leqslant 10$ and $x_1 + 2x_2 \leqslant 15$. There should be one row between the x's and the line for each nontrivial constraint. Under x_1 and x_2 in the first row are the coefficients for $x_1 + x_2 \leqslant 10$. The 10 is under the constant label. Under that row are the constants for $x_1 + 2x_2 \leqslant 15$. The 15 is under the constant label. Also a $\begin{smallmatrix}1&0\\0&1\end{smallmatrix}$ diagonal matrix is placed underneath the artificial variables. And the artificial variables x_3 and x_4 are written in a column to the extreme right with zeroes to their right. Row 1 starts out with all zeros, and the P_i's-Row 1 is just that.

To get to Table 2, the P_i's row – Row 1 is surveyed to find the largest positive number (in this case 5). Therefore, we will "pivot" in this column. The row we pivot in is determined by taking the numbers in that column (between the x's and the line) and dividing them into the constant term in their individual row. So 1 into 10 equals 10, and 2 into 15 equals 7.5. The 7.5 is smaller, so we take the row that produced it and pivot there; note the circled 2. At this point we use Rule 2 of solving systems of equations on that row to turn the 2 into a 1. In other words, we divide that row by 2. Then, in moving to Table 3 we use Rule 3 from solving systems of equations to put a zero in the slot above the pivot element. To do this –1 times, the bottom row is added to the top row. And this sum becomes the new top row. The bottom row stays the same. At this point, we pivot the 5 and x_2 heading the pivot column into the extreme right-hand part of the pivot row (replacing x_4 and 0). Now the new Row 1 is determined by multiplying the numbers under the coefficient column times each number in its row and then adding that product to the similar product from the other row (column by column). So $0 \times 1/2 + 5 \times 1/2 = 5/2$. And $0 \times 0 + 5 \times 1 = 5$. And $0 \times 1 + 5 \times 0 = 0$. And $0 \times (-1/2) + 5 \times 1/2 = 5/2$. And $0 \times 5/2 + 5 \times 15/2 = 75/2$.

Then the P_i's-Row 1 is calculated. At this point the largest positive value is 3/2 (the only one), so we pivot in column 1 and repeat the entire Table 1 through Table 3 process in Table 3 through Table 5.

We pivot at the circled 1/2, because 1/2 into 5/2 equals 5 is less than 1/2 into 15/2 equals 15.

Then in Table 4, the pivot row is multiplied by 2 to turn the 1/2 into a 1 (Rule 2 of systems of equations). Then –1/2 times that row is added to the bottom row to form the new bottom row. Then the x_1 and 4 heading that column are moved to the extreme right-hand side of the pivot row to replace x_3 and 0.

Then Row 1 is calculated. 4 x 1 + 5 x 0 = 4. 4 x 0 + 5 x 1 = 5. 4 x 2 + 5 x (−1) = 3. 4 x (−1) + 5 x (1) = 1, and 4 x 5 + 5 x 5 = 45. Then the P_i's-Row 1 row is calculated. Because there are no positive numbers in this row, we are done and the solution is x_1 and x_2 equal the values in their constant columns and P is the value in the constant column and Row 1 row. So $x_1 = 5, x_2 = 5$ with $P = 45$.

EXAMPLE 11.3

Let us do another simplex example. Maximize $P = 10x_1 + x_2$ subject to $x_1 \geqslant 0, x_2 \geqslant 0, x_1 + 3x_2 \leqslant 16$, and $3x_1$ and $x_2 \leqslant 16$.

Table 1

	x_1	x_2	x_3	x_4	constant		coefficient
P_i's	10	1	0	0			
	1	3	1	0	16	x_3	0
	(3)	1	0	1	16	x_4	0
Row 1	0	0	0	0	0		
P_i's-Row 1	10	1	0	0			

Table 2

x_1	x_2	x_3	x_4	constant
1	3	1	0	16
1	1/3	0	1/3	16/3

Table 3

	x_1	x_2	x_3	x_4	constant		coefficient
	10	1	0	0			
	0	8/3	1	−1/3	32/3	x_3	0
	1	1/3	0	1/3	16/3	x_1	10
Row 1	10	10/3	0	10/3	160/3		
P_i's-Row 1	0	−7/3	0	−10/3	so done		

The answer is $x_1 = 16/3$, $x_2 = 0$ with a corresponding $P = 160/3$.

This time everything is set up as before except that at Table 3 in P_i's-Row 1 there are no positive values, so we are done and the answer is again in the constant column. Any artificial variables (in this case x_3) in the constraint column are not part of the solution and any real variables that do not make it into the constant column array are zero. So in this case $x_1 = 16/3, x_2 = 0$ yields the maximum P value of 160/3.

Notice in these simplex problems, that a large problem with, say, 10 constraints would require ten rows between the x's and the line and a 10 by 10 diagonal matrix with 1's down the main diagonal and 0's everywhere else. At each pivot, all other nine numbers in the pivot column would have to be turned into 0's using Rule 3, etc. Consequently, pencil and paper calculation would be enormous and the chances for error substantial. Computer programs have been written that do the simplex calculations automatically. They are sold by software companies and computer manufacturers. These programs are recommended for large-scale and/or frequent LP applications.

However, for a five, ten, fifteen or maybe even twenty variable problems, one could ignore the simplex technique and use our simulation technique. This could be recommended for the occasional user of linear programming or one who does mostly nonlinear applications.

To that end, we present a small discussion of hints for solving problems without using the simplex technique. The reader should be aware that there are LP books which explain and prove the simplex algorithm for the serious mathematician who is interested in more rigor than this explanation provides. Also, the simplex technique can be modified to solve minimization problems, and, of course, the minimum is at a "corner" point, too. Also multistage (chapter 12) works on many LP problems.

IMPROVING THE EFFICIENCY OF THE PROGRAMS.

EXAMPLE 11.4

Consider the problem to maximize $P = 7x^2 - xy + 3y^2 + xy^2$ subject to $x \geqslant 0$, $y \geqslant 0$, and $x + y \leqslant 5000$. The constraints tell us that $0 \leqslant x \leqslant 5000$ and $0 \leqslant y \leqslant 5000$. Therefore, there are 5001 x 5001 = 25,010,001 combinations to try. However, we could exploit the fact that as x ranges from 0 to 5000, y must range from 0 to $5000 - x$. The program would look like this:

```
10   B=-999999
20   FOR X=0 TO 5000
30   K=5000-X
40   FOR Y=0 TO K
50   P=7*X**2-X*Y+3*Y**2+X*Y**2
60   IF P >  B THEN 80
70   GO TO 110
80   A1=X
```

```
90   A2=Y
100  B=P
110  NEXT Y
120  NEXT X
130  PRINT A1,A2,B
140  STOP
150  END
```

If this approach is tried on a Monte Carlo simulation, then you must multiply the second random number by $5000 - x + 1$ which will guarantee that $x + y \leqslant 5000$ (still referring to the previous inequality).

If the constraint were $x + y + z + w \leqslant 5000$, then multiply the first random number by 5001. Multiply the second random number by $5001 - x$. Multiply the third random number by $5001 - x - y$. Multiply the fourth random number by $5001 - x - y - z$. This could introduce a slight bias in the random process. However, it should not cause much of a problem in practice, and there are other "bias free" approaches to the random process.

One "bias free" approach is to read in four random numbers between 0 and 5000 (referring to $x + y + z + w \leqslant 5000$) and add them up. If the sum is less than or equal to 5000, use them alone. If the sum is greater than 5000, divide 5000 into the sum and divide each random number by that quotient. This will "shrink" the random numbers small enough so that $x + y + z + w \leqslant 5000$.

There are other approaches for more sophisticated problems. The question is, though, is it worth the extra programming trouble? The author thinks the answer is sometimes yes.

EXAMPLE 11.5

Assume that a function of three variables is constrained by $1 \leqslant x \leqslant 1000$, $1 \leqslant y \leqslant 1000$, and $1 \leqslant z \leqslant 1000$. Therefore, there are 1000 x 1000 x 1000 = 1,000,000,000 possible answers. If this is too many to check on your computer system, consider letting x, y, and z increase by increments of ten instead of one. Therefore, x could take the values 10, 20, 30, 40, . . . , 1000; y and z could do the same. This way there would be only 100 x 100 x 100 = 1,000,000 possibilities to check.

This would probably work well in business where there is frequently 10% to 20% estimation error in the model to begin with.

EXAMPLE 11.6

Consider the problem to maximize $P = 20x_1 + 15x_2 + 19x_3 + 27x_4$

$+ 34x_5 + 42x_6 + 58x_7 + 21x_8 + 90x_9 + 66x_{10} + 15x_{11} + 75x_{12} + 14x_{13} + 88x_{14} + 62x_{15} + 60x_{16} + 58x_{17} + 54x_{18} + 90x_{19} + 29x_{20}$ subject to $x_i \geqslant 0$ for all i, and $4x_1 + 5x_2 + 2x_3 + 2x_4 + x_5 + 5x_6 + 6x_7 + 5x_8 + 4x_9 + 3x_{10} + 5x_{11} + 6x_{12} + 2x_{13} + 8x_{14} + 6x_{15} + 5x_{16} + x_{17} + x_{18} + 5x_{19} + 5x_{20} \leqslant 3800$, $x_1 + x_2 + 8x_3 + 6x_4 + 4x_5 + 2x_6 + 3x_7 + 2x_8 + 4x_9 + 6x_{10} + x_{11} + 2x_{12} + x_{13} + 2x_{14} + 6x_{15} + x_{16} + 3x_{17} + 4x_{18} + 2x_{19} + 5x_{20} \leqslant 3800$, $3x_1 + 2x_2 + x_3 + 2x_4 + x_5 + x_6 + 3x_7 + x_8 + x_9 + 2x_{10} + 2x_{11} + 2x_{12} + 5x_{13} + x_{14} + x_{15} + x_{16} + 6x_{17} + x_{18} + 5x_{19} + 2x_{20} \leqslant 3800$, and $2x_1 + 2x_2 + 2x_3 + 3x_4 + 2x_5 + 2x_6 + x_7 + 4x_8 + x_9 + x_{10} + 3x_{11} + 2x_{12} + 7x_{13} + 6x_{14} + 2x_{15} + x_{16} + 7x_{17} + 5x_{18} + 3x_{19} + 4x_{20} \leqslant 3800$.

This is a straight linear programming problem which could be solved by the simplex method if a simplex package program were obtainable. Also, we could produce a simulated optimum which would be fairly close to the true optimum. However, let us use our simulation technique along with a little linear programming theory to produce the optimum without resorting to a simplex package program.

First we write a program that reads in twenty random numbers (each time through the loop) and "shrinks" them back inside the constraints as discussed previously. There are no biases this way. After a few runs of the simulation, the printouts begin to identify certain key variables that tend to have large values, while the others tend to be around zero (relatively speaking).

It is a well-known fact of linear programming theory that in a maximization problem, if all the inequalities are of the less than type, then the number of nonzero variables in the optimum solution cannot exceed the number of inequalities. In this case there are no more than four nonzero variables. Let us use this fact.

The author noticed six promising (never near zero) variables from the twenty-variable simulation runs. Therefore, the program was modified so that only six variables were nonzero in the simulation.

Then several more runs were made. The printouts are given below:

```
      .000       .000       .000       .000     2.079
      .000       .000       .000    460.552     5.451
      .000       .000       .000       .000      .000
      .000    232.524    134.300    314.508      .000
90924.313
```

```
       .000      .000      .000      .000   36.611
       .000      .000      .000   482.270    6.769
       .000      .000      .000      .000     .000
       .000   287.899    58.351   293.384     .000
  91349.313
       .000      .000      .000      .000    5.823
       .000      .000      .000   468.716   21.681
       .000      .000      .000      .000     .000
       .000   252.530   104.870   297.347     .000
  90884.250
       .000      .000      .000      .000   14.452
       .000      .000      .000   557.580   25.045
       .000      .000      .000      .000     .000
       .000   292.447     1.967   237.136     .000
  90736.813
```

From the printouts it looks like x_9, x_{17}, x_{18}, and x_{19} are the nonzero variables. This reduces the model to maximize $P = 90x_9 + 58x_{17} + 54x_{18} + 90x_{19}$ subject to $x_i \geqslant 0$, $4x_9 + x_{17} + x_{18} + 5x_{19} \leqslant 3800$, $4x_9 + 3x_{17} + 4x_{18} + 2x_{19} \leqslant 3800$, $x_9 + 6x_{17} + x_{18} + 5x_{19} \leqslant 3800$, and $x_9 + 7x_{17} + 5x_{18} + 3x_{19} \leqslant 3800$. This is easily solvable through another simulation.

TRANSPORTATION PROBLEMS

The classic transportation problem is merely a special case of LP and so may be solved by a slight modification of the simplex technique. Most package programs that solve LP problems will also work on transportation problems.

EXAMPLE 11.7

Let us state an example of a transportation problem. The Idaho Company has two factories and three warehouses with the following supplies and demands and unit shipping costs for Product R.

	Warehouse 1	Warehouse 2	Warehouse 3	Supply
Factory 1	2.10	5.16	3.23	100
Factory 2	1.59	2.68	3.10	300
	150	175	75	

It costs \$2.10 to ship one unit from Factory 1 to Warehouse 1 and \$3.10 to ship one unit from Factory 2 to Warehouse 3, for example. Also, Factory 1 has 100 units to ship, Factory 2 has 300 units to ship, and warehouses 1, 2, and 3 need 150, 175, and 75 units, respectively. Therefore, the problem is to minimize $C = 2.10x_{11} + 5.16x_{12} + 3.23x_{13} + 1.59x_{21} + 2.68x_{22} + 3.10x_{23}$ subject to $x_{11} \geqslant 0, x_{12} \geqslant 0, x_{13} \geqslant 0, x_{21} \geqslant 0, x_{22} \geqslant 0, x_{23} \geqslant 0, x_{11} + x_{12} + x_{13} = 100, x_{21} + x_{22} + x_{23} = 300, x_{11} + x_{21} = 150, x_{12} + x_{22} = 175$, and $x_{13} + x_{23} = 75$, where x_{ij} represents the number of units of Product R shipped from factory i to warehouse j.

Some of the difficulties involved in applying transportation problems are the unit costs assumptions and the fact that most transportation problems involve more than one product. But even with these and other difficulties, businesses have saved millions of dollars using the transportation model.

Note that in the packaging problems earlier in the book, we tried to attack the transportation problem by reducing the unit shipping cost, by getting more containers in a boxcar, truck, ship, etc.

Either/or both approaches could be taken, depending upon the circumstances.

EXERCISES

11.1 Maximize $P = 3x + 7y$ subject to $x \geqslant 0, y \geqslant 0, x + 3y \leqslant 30$, and $x + y \leqslant 20$, using graphical techniques.

11.2 Solve exercise 11.1 using the simplex technique.

11.3 Solve exercise 11.1 using our search all techniques by writing a BASIC program as discussed throughout the book.

11.4 Minimize $C = 6x + 4y$ subject to $x \geqslant 0, y \geqslant 0, x + 2y \leqslant 15$, and $x + y \leqslant 10$, using the graphical technique.

11.5 Try to write a computer program to automate solving any LP program by the simplex technique. (Hint: This may be quite hard.)

11.6 The Marquette Company has the following supplies, demands, and unit shipping costs for its product:

	WH 1	WH2	WH3	
Factory 1	2.18	9.14	3.92	750
Factory 2	7.12	3.85	6.50	550
Factory 3	4.53	12.16	7.20	700
	500	900	600	

State this transportation problem in algebraic terms.

11.7 The Swedenborg Company is manufacturing products A, B, and C in departments 1 and 2. Product A returns \$6 per unit profit, B returns \$12 per unit, and C returns \$10. It takes 1 hour in Department 1 and 1 hour in Department 2 to make one unit of A. It takes 2 hours in Department 1 and 1 hour in Department 2 to make one unit of B. And it takes 1.5 hours in each department to make one unit of C. There are 20,000 worker-hours available per week in Department 1 and 15,000 worker-hours available per week in Department 2. State this maximize LP problem algebraically and then solve it using the simplex technique.

11.8 Write a BASIC program to maximize Swedenborg's profit (exercise 11.7) by searching all the solutions.

11.9 Maximize $P = 6x_1 + 5x_2 + 10x_3$ subject to $x_1 \geqslant 0, x_2 \geqslant 0, x_3 \geqslant 0, x_1 + x_2 + 2x_3 \leqslant 1000, 2x_1 + 3x_2 + 5x_3 \leqslant 25000$, using the simplex technique.

11.10 The Negaunee-Manistique Company has three factories and three warehouses. They are planning to build a fourth warehouse at either their Red Cedar or Pecos location. These factories supply and ship their Grade A tires. The unit shipping costs and estimated supplies and demands for each warehouse possibility (along with the ones already built) are below:

	WH 1	WH 2	WH 3	Red Cedar WH	Supplies
Factory 1	\$1.02	\$ .62	\$ .75	\$ .48	18,000
Factory 2	\$ .79	\$ .71	\$1.03	\$ .29	20,000
Factory 3	\$ .91	\$1.09	\$ .84	\$ .58	12,000
	10,000	20,000	12,000	8,000	

Warehouse Demands

	WH 1	WH 2	WH 3	Pecos WH	
Factory 1	$1.02	$ .62	$ .75	$ .42	18,000
Factory 2	$ .79	$ .71	$1.03	$ .45	20,000
Factory 3	$ .91	$1.09	$ .84	$ .51	12,000
	10,000	20,000	12,000	8,000	

All other costs being equal, should Negaunee-Manistique build at Red Cedar or Pecos? Just state the problem.

11.11 The Lutsen Feed Company blends two types of feed, A and B, according to the following chart:

	Ingredient 1	Ingredient 2	Ingredient 3
Feed A	At least 25%	No more than 40%	Any amount
Feed B	No more than 30%	Any amount	At least 33%
	18% Nut. 1 5% Nut. 2 In Ing. 1	12% Nut. 1 16% Nut. 2 In Ing. 2	16% Nut. 1 15% Nut. 2 In Ing. 3

Feed A must have exactly 20% of Nutrient 1 and exactly 15% of Nutrient 2. Feed B must have exactly 15% of Nutrient 1 and exactly 15% of Nutrient 2.

Feed A sells for $20 a bag, Feed B sells for $17 a bag. Ingredient 1 costs $5 a bag, Ingredient 2 costs $7 a bag, and Ingredient 3 costs $6 a bag. (All five bag sizes are the same.) There are 15,000 bags of each type of ingredient available, and Lutsen can purchase any or all of it. Find the maximum profit solution. Just state the question. This would be a lot of work to solve using the simplex technique with pencil and paper. However, using our simulation or search all techniques, the solution is fairly straightforward.

Suggested Reading

Bellman, R. and Dreyfus, S. *Applied Dynamic Programming.* Princeton, N.J.: Princeton University Press, 1962.

Dantzig, G.B. *Linear Programming and Its Extensions.* Princeton, N.J.: Princeton University Press, 1963.

Gass, S.I. *Linear Programming: Methods and Applications,* 3d ed. New York: McGraw-Hill, 1969.

Hadley, G. *Linear Programming.* Reading, Mass.: Addison-Wesley, 1962.

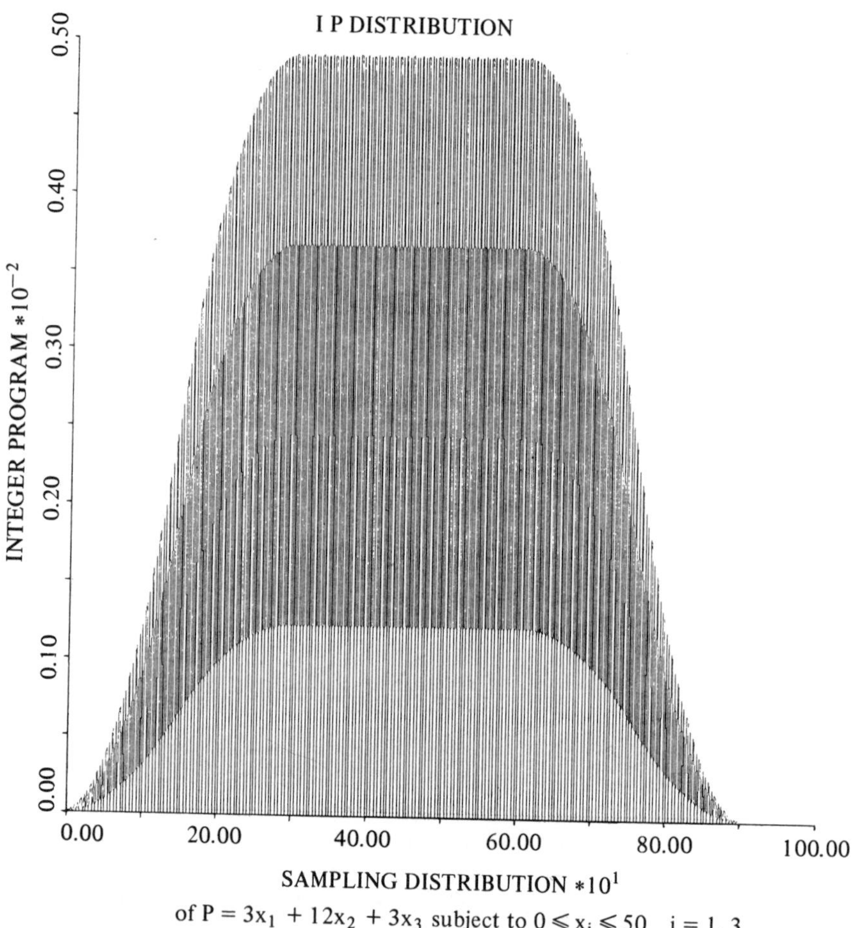

FIGURE 12.1. The multistage Monte Carlo integer programs in this chapter are the ones that travel across the distribution of all answers until they funnel into the correct answer. At first it seemed as though this technique would work for only a few variables. However, after experimenting, it appears that programs with up to 100 variables or more can be optimized (at least find the approximate optimum). The true optimum can be found easily for up to twenty variables. This makes the solution of nonlinear models a reality. (Reprinted by permission of the National Bureau of Standards, Special Publication 503, p. 365.)

CHAPTER 12

Multistage Monte Carlo Integer Programs

Throughout the text we have seen how to run multiple Monte Carlo simulations on the same problem to discover the area in which the true optimum lies. Then we have proceeded to "focus" in on this area with more simulations or at that point perform an exhaustive search for the true optimum. This is usually not too demanding and is recommended for many problems.

However, it seems like one should be able to automate that process and let the computer focus into the right region and find the true optimum. This would free the problem solver for other tasks while this process was going on. Also, if it can be done successfully, it could greatly reduce simulation time and pave the way for solving large-scale nonlinear optimization problems. These are the goals of this chapter.

EXAMPLE 12.1

Maximize $P = 350{,}000 - x_1^2 + 600x_1 - x_2^2 + 800x_2 - x_3^2 + 1000x_3 - x_4^2 + 1200x_4 - x_5^2 + 1400x_5$ subject to $0 \leqslant x_i \leqslant 1000$ for $i = 1.5$.

There are only five variables and $1001^5 = 1.00501001 \times 10^{15}$ answers, so repeated simulations followed by a "focus" search would probably yield the optimum. Also, we could factor the right side of the equation as follows to give us maximize $P = 1{,}000{,}000 - (x_1 - 300)^2 - (x_2 - 400)^2 - (x_3 - 500)^2 - (x_4 - 600)^2 - (x_5 - 700)^2$, $0 \leqslant x \leqslant 1000$, $i = 1{,}5$. At this point we can see that $x_1 = 300$, $x_2 = 400$, $x_3 = 500$, $x_4 = 600$, $x_5 = 700$ yields a maximum P of 1,000,000.

However, let us write a multistage Monte Carlo integer program that automatically keeps focusing in the right direction toward the true but unknown optimum. The program is presented below with the explanation following. Notice that we are using a problem that we already know the answer to, just to verify that this approach works. Any function can be substituted in place of this example and the program will find the optimal solution. This program takes less than one minute. Notice the printout below the program. The true optimum was found long before the completion of the computer run.

```
5     DIM A(5),B(5),L(5),N(5),U(5),X(5),P(5),S(5)
7     F=2
10    X=1
11    S(1)=300, S(2)=400, S(3)=500, S(4)=600, S(5)=700
12    M=900000
14    A(1)=500, A(2)=500, A(3)=500, A(4)=500,A(5)=500
16    B(1)=0, B(2)=0, B(3)=0, B(4)=0, B(5)=0
18    N(1)=1001, N(2)=1001, N(3)=1001, N(4)=1001,
        N(5)=1001
20    FOR J=1 TO 9
22    FOR I=1 TO 5000
24    FOR K=1 TO 5
30    IF A(K)-N(K)/F**J < B(K) THEN 50
40    GO TO 60
50    L(K)=B(K)
55    GO TO 65
60    L(K)=A(K)-N(K)/F**J
65    IF A(K)+N(K)/F**J > N(K) THEN 80
70    GO TO 90
80    U(K)=N(K)-L(K)
85    GO TO 100
90    U(K)=A(K)+N(K)/F**J-L(K)
100   X(K)=INT(L(K)+RND(X)*U(K))
102   NEXT K
105   P=1000000
110   FOR V=1 TO 5
120   P=P-(X(V)-S(V))**2
130   NEXT V
140   IF P > M THEN 160
150   GO TO 170
160   A(1)=X(1), A(2)=X(2), A(3)=X(3), A(4)=X(4),
        A(5)=X(5), M=P
170   NEXT I
175   PRINT A(1),A(2),A(3),A(4),A(5),M
180   NEXT J
```

```
190  PRINT A(1),A(2),A(3),A(4),A(5)
200  PRINT M
210  STOP
220  END

339            464           464           501           680
982886
337            392           543           619           673
995628
305            420           499           595           678
999065
298            390           496           595           698
999851
299            402           495           596           698
999950
299            402           498           602           701
999986
300            400           500           600           700
1000000
300            400           500           600           700
1000000
300            400           500           600           700
1000000
300            400           500           600           700
1000000
```

EXPLANATION OF THE PROGRAM

Nine consecutive simulations of 5000 feasible solutions are done over an ever narrowing range. The center point of the ever narrowing range of solutions is always the best optimal answer so far. So the program quite naturally just funnels into the right answer.

Because each x_i is between 0 and 1000, we selected $x_1 = 500$, $x_2 = 500$, $x_3 = 500$, $x_4 = 500$, $x_5 = 500$ with a corresponding $P = 900{,}000$ as the first answer. See lines 12 and 14. This is put in the storage area line 160 and will remain there until a better answer comes along. This is just to get the program started.

Line 5 declares a few subscripted variables for the program. Line 7 sets F = 2; F will be the factor by which we decrease the size of the variable range each time. When J=1 (line 20), each x_i will have a range of 1000 (0 to 1000). When J=2, each variable will have a range of at most 500 (plus or minus 250 around the individual x_i coordinates in the optimal solution so far). The range could be less than 250 in one direction if the x_i in question is near a boundary (in this case 0 or 1000). Then that range limit will be the boundary value of 0 or 1000.

When J=3, each variable will have a range of at most 250 (plus or minus 125). When J=4, each variable will have a range of at most 125. These range adjustments are taking place in lines 30, 40, 50, 55, 60, 65, 70, 80, 85, and 90. When J=5, each variable will have a range of at most 62. When J=6, the range will be at most 31. With J=7, the range at most will be 15. J=8 will produce a range of 7, and J=9 will produce a range of about 3. This last simulation of 5000 in such a narrow range will produce the true optimum.

This is actually a very conservative program. The optimum could have been produced with less than the 5000 samples on each of the 9 loops.

Line 10 (x=1) just gets the random number generator RND(x) started. Line 11 puts in the factors 300, 400, 500, 600, and 700 in the second (factored) version of our objective function. The function is in lines 105, 110, 120, and 130. This program in no way exploits the fact that we already know the answer. The author just chose this factored form (line 120) because it was easy to type. Line 16 sets the lower bounds for the x_i's, while line 18 sets the upper bounds for the x_i's.

Line 20 and 180 form the outer loops of nine increasingly focused searches about the optimum so far. Lines 22 and 170 form the inner loop which is done 5000 times for each of the nine outer loop runs. Notice that after each inner run of 5000, the optimum so far is printed from line 175.

Lines 24 and 102 set up the loop that produces the five random numbers for each solution. This loop also controls (and subsequently narrows) the range about the optimal x_i's so far. In this area L(K) becomes the lower bound of the range and U(K) becomes its width so that line 100 can produce a random number in the range. Then the function is evaluated (lines 105, 110, 120, and 130) for the five random x_i values. In line 140 the current P value is checked to see if it is larger than the optimum so far, M. If it is, then it goes to line 160 and stores the new optimal solution. If not, the program follows GO TO 170 and goes on to the next randomly generated solution.

Notice in the lines 24 through 102 loop, line 30 checks to see if the bottom of the range is less than B(K)=0. If it is, then L(K), the bottom of the range in line 100, becomes B(K)=0 in this case. If not, then L(K) becomes A(K)–N(K)/F**J which is the natural lower end (bottom) of the range. A(K) is the x_k value of the optimum so far. N(K) is the upper bound on x_k. F**J is 2**J in this case, or how much the range is to be shrunk by for given J.

Line 65 checks to see if the upper end (top) of the range is out of the range of the variable. If it is, then line 80 makes U(K)=N(K)–L(K) where U(K) is the width of the range for line 100. If not, then line 90 makes U(K) what the width of the range would be if the upper bound or x_k does not interfere. All of this gets adjusted automatically every time a new optimum so far is produced and every time J gets incremented in the outer loop.

As you can see from the printout, the computer is on the right track to the optimum after the first simulation 339, 464, 464, 501, 680. After that it just keeps zeroing in on it until it has it.

EXAMPLE 12.2

Now let us look at another example. Maximize $P = 407{,}200{,}000{,}000 - (x_1 - 265{,}318)^2 - (x_2 - 412{,}400)^2 - (x_3 - 97{,}000)^2 - (x_4 - 914{,}002)^2 - (x_5 - 600{,}000)^2$ subject to $0 \leqslant x_i \leqslant 1{,}000{,}000$.

There are 1.000005×10^{30} answers. If we set up five nested loops (from 0 to 1,000,000) and looked at all 1.000005×10^{30} answers (to get the true optimum), it would take about 32 million trillion years of computer time on a medium-sized computer. So we will substitute this function to be maximized into our multistage Monte Carlo integer program and let it find the true optimum as before. The only change is in line 20 where we have the outer loop run 19 times, because we are narrowing a range of 1,000,000 instead of 1000 from before.

Notice from our factored form that the true maximum is at $x_1 = 265{,}318$, $x_2 = 412{,}400$, $x_3 = 97{,}000$, $x_4 = 914{,}002$, and $x_5 = 600{,}000$. As you can see from looking at the printout, the program had no trouble finding the exact optimum.

```
5    DIM A(5), B(5), L(5), N(5), U(5), X(5), P(5), S(5)
7    F=2
10   X=1
11   S(1)=265318, S(2)=412400, S(3)=97000, S(4)=914002,
       S(5)=600000
12   M=643942900
14   A(1)=500000, A(2)=500000, A(3)=500000, A(4)=500000
       A(5)=500000
16   B(1)=0, B(2)=0, B(3)=0, B(4)=0, B(5)=0
18   N(1)=1000001, N(2)=1000001, N(3)=1000001,
       N(4)=1000001, N(5)=1000001
20   FOR J=1 TO 19
```

```
22    FOR I=1 TO 5000
24    FOR K=1 TO 5
30    IF A(K)-N(K)/F**J < B(K) THEN 50
40    GO TO 60
50    L(K)=B(K)
55    GO TO 65
60    L(K)=A(K)-N(K)/F**J
65    IF A(K)+N(K)/F**J > N(K) THEN 80
70    GO TO 90
80    U(K)=N(K)-L(K)
85    GO TO 100
90    U(K)=A(K)+N(K)/F**J-L(K)
100   X(K)=INT(L(K)+RND(X)*U(K))
102   NEXT K
105   P=40.72E*10
110   FOR V=1 TO 5
120   P=P-(X(V)-S(V))**2
130   NEXT V
140   IF P > M THEN 160
150   GO TO 170
160   A(1)=X(1),A(2)=X(2),A(3)=X(3),A(4)=X(4),
        A(5)=X(5), M=P
170   NEXT I
175   PRINT A(1), A(2), A(3), A(4), A(5), M
180   NEXT J
190   PRINT A(1), A(2), A(3), A(4), A(5)
200   PRINT M
210   STOP
220   END
```

```
215532    475910    42018     821839    597922
3.89166E+11
290208    443550    112824    937912    609882
4.04690E+11
273186    410010    100443    911047    623856
4.06543E+11
274925    413752    107240    918153    605314
4.06956E+11
264774    416309    88930     910270    597798
4.07101E+11
```

267786	412103	99080	915247	598805
4.07187E+11				
265629	411176	97291	912850	600515
4.07197E+11				
265031	411492	96735	914483	599675
4.07199E+11				
264991	412133	97019	914311	599986
4.07200E+11				
265344	412317	96886	913795	600034
4.07200E+11				
265262	412379	97012	913980	600007
4.07200E+11				
265317	412381	97013	913973	599948
4.07200E+11				
265319	412423	97000	914005	600011
4.07200E+11				
265321	412388	97007	913995	600001
4.07200E+11				
265321	412400	97007	913998	600002
4.07200E+11				
265317	412403	97000	914003	600002
4.07200E+11				
265317	412400	96999	914001	600001
4.07200E+11				
265318	412400	97000	914002	600000
4.07200E+11				
265318	412400	97000	914002	600000
4.07200E+11				
4.07200E+11				

EXAMPLE 12.3

Let us try to maximize $P = 407{,}200{,}000{,}000 - (x_1 - 265{,}318)^2 - (x_2 - 412{,}400)^2 - (x_3 - 97{,}000)^2 - (x_4 - 914{,}002)^2 - (x_5 - 600{,}000)^2 + x_2{}^{1.6} + x_2 x_1{}^{.6} + x_2 x_3{}^{.6} + x_2 x_4{}^{.6} + x_2 x_5{}^{.6}$ subject to $0 \leqslant x_i \leqslant 1{,}000{,}000$.

This time because of the cross-product terms the optimum is unknown. However, we merely substitute the function into our multistage program and let it run for a minute or two. The resulting maximum is $x_1 = 266{,}168$, $x_2 = 419{,}037$, $x_3 = 98{,}266$, $x_4 = 914{,}521$, $x_5 = 600{,}614$ with a corresponding maximum of P of 4.12121×10^{11}. See the following program and printout:

```
5     DIM A(5), B(5), L(5), N(5), U(5), X(5), P(5), S(5)
7     F=2
10    X=1
11    S(1)=265318, S(2)=412400, S(3)=97000, S(4)=914002,
       S(5)=600000
12    M=678436141.5
14    A(1)=500000, A(2)=500000, A(3)=500000,
         A(4)=500000, A(5)=500000
16    B(1)=0, B(2)=0, B(3)=0, B(4)=0, B(5)=0
18    N(1)=1000001,N(2)=1000001, N(3)=1000001,
         N(4)=1000001, N(5)=1000001
20    FOR J=1 TO 19
22    FOR I=1 TO 5000
24    FOR K=1 TO 5
30    IF A(K)-N(K)/F**J < B(K) THEN 50
40    GO TO 60
50    L(K)=B(K)
55    GO TO 65
60    L(K)=A(K) -N(K)/F**J
65    IF A(K)+N(K)/F**J >  N(K) THEN 80
70    GO TO 90
80    U(K)=N(K)-L(K)
85    GO TO 100
90    U(K)=A(K)+N(K)/F**J-L(K)
100   X(K)=INT(L(K)+RND(X)*U(K))
102   NEXT K
105   P=40.72E+10
110   FOR V=1 TO 5
120   P=P-(X(V)-S(V))**2+X(2)*X(V)**.6
130   NEXT V
140   IF P  > M THEN 160
150   GO TO 170
160   A(1)=X(1), A(2)=X(2), A(3)=X(3), A(4)=X(4),
         A(5)=X(5), M=P
```

```
170  NEXT I
175  PRINT A(1), A(2), A(3), A(4), A(5), M
180  NEXT J
190  PRINT A(1), A(2), A(3), A(4), A(5)
200  PRINT M
210  STOP
220  END
```

```
217238      361733   65219    837588   550050
3.96916E+11
236065      426829   89037    297096   602825
4.10903E+11
236065      426829   89037    927096   602825
4.10903E+11
254595      412967   89088    927162   602762
4.11702E+11
264087      417499   93334    911840   597590
4.12074E+11
266524      421272   94024    913269   599947
4.129096E+11
267758      418549   97819    913665   599401
4.12116E+11
266169      418886   98499    914711   599546
4.12120E+11
266559      418802   98382    914759   600865
4.12121E+11
266026      418959   98260    914543   600666
4.12121E+11
266103      419079   98283    914576   600650
4.12121E+11
266152      419066   98256    914518   600546
4.12121E+11
```

266186	419041	98264	914539	600598
4.12121E+11				
266165	419031	98269	914532	600613
4.12121E+11				
266168	419035	98263	914514	600610
4.12121E+11				
266171	419035	98267	914522	600615
4.12121E+11				
266167	419038	98267	914521	600614
4.12121E+11				
266167	419037	98266	914521	600614
4.12121E+11				
266168	419037	98266	914521	600614
4.12121E+11				
266168	419037	98266	914521	600614
4.12121E+11				

EXAMPLE 12.4

Consider the following problem: maximize $P = 1{,}914{,}258 - (x_1 - 752)^2 - (x_2 - 914)^2 - (x_3 - 212)^2 - (x_4 - 444)^2 - (x_5 - 14)^2 + x_3x_1 + x_3x_2 + x_3^2 + x_3x_4 + x_3x_5$ subject to $0 \leqslant x_i \leqslant 1000$ for $i = 1{,}5$.

Again, because of the cross-product terms the optimum is unknown. We merely substitute the function into our multistage program and let it run for a minute or two. The resulting maximum is $x_1 = 1000$, $x_2 = 1000$, $x_3 = 1000$, $x_4 = 944$, $x_5 = 514$ with a corresponding maximum of P of 5,182,414. See the program and printout below:

```
5    DIM A(5), B(5), L(5), N(5), U(5), X(5), P(5), S(5)
7    F=2
10   X=1
11   S(1)=752, S(2)=914, S(3)=212, S(4)=444, S(5)=14
12   M=2607082
14   A(1)=500, A(2)=500, A(3)=500, A(4)=500, A(5)=500
16   B(1)=0, B(2)=0, B(3)=0, B(4)=0, B(5)=0
18   N(1)=1001, N(2)=1001, N(3)=1001, N(4)=1001, N(5)=1001
```

```
20    FOR J=1 TO 9
22    FOR I=1 TO 5000
24    FOR K=1 TO 5
30    IF A(K)-N(K)/F**J  <  B(K) THEN 50
40    GO TO 60
50    L(K)=B(K)
55    GO TO 65
60    L(K)=A(K)-N(K)/F**J
65    IF A(K)+N(K)/F**J >  N(K) THEN 80
70    GO TO 90
80    U(K)=N(K)-L(K)
85    GO TO 100
90    U(K)=A(K)+N(K)/F**J-L(K)
100   X(K)=INT(L(K)+RND(X)*U(K))
102   NEXT K
105   P=1914258
110   FOR V=1 TO 5
120   P=P-(X(V)-S(V))**2+X(3)*X(V)
130   NEXT V
140   IF P > M THEN 160
150   GO TO 170
160   A(1)=X(1), A(2)=X(2), A(3)=X(3), A(4)=X(4),
        A(5)=X(5), M=P
170   NEXT I
175   PRINT A(1), A(2), A(3), A(4), A(5), M
180   NEXT J
190   PRINT A(1), A(2), A(3), A(4), A(5)
200   PRINT M
210   STOP
220   END
```

```
906       951    995     841   506
5054416
987       944    992     936   477
5094612
994       992    999     928   472
5166836
988       993    1000    967   533
5169487
1000      999    1000    915   571
5177495
```

```
1000      1000   1000   926   563
5179689
1000      1000   1000   931   521
5182196
1000      1000   1000   944   515
5182413
1000      1000   1000   944   514
5182414
1000      1000   1000   944   514
5182414
```

EXAMPLE 12.5

Let us maximize $P = 1{,}914{,}258 - (x_1 - 752)^2 - (x_2 - 914)^2 - (x_3 - 212)^2 - (x_4 - 444)^2 - (x_5 - 14)^2 + x_3 x_1{}^{.8} + x_3 x_2{}^{.8} + x_3 x_3{}^{.8} + x_3 x_4{}^{.8} + x_3 x_5{}^{.8}$ subject to $0 \leqslant x_i \leqslant 1000$ for $i = 1{,}5$.

Here, too, the optimum is unknown because of the cross-product terms. But after substituting the function into the multistage program we get the true optimum $x_1 = 827$, $x_2 = 986$, $x_3 = 716$, $x_4 = 526$, $x_5 = 123$ with a corresponding $P = 2.24204 \times 10^6$. See the program and printout below:

```
5        DIM A(5),B(5),L(5),N(5),U(5),X(5),P(5),S(5)
7        F=2
10       X=1
11       S(1)=752,S(2)=914,S(3)=212,S(4)=444,S(5)=14
12       M=1461151
14       A(1)=500,A(2)=500,A(3)=500,A(4)=500,A(5)=500
16       B(1)=0,B(2)=0, B(3)=0, B(4)=0, B(5)=0
18       N(1)=1001,N(2)=1001,N(3)=1001,N(4)=1001,N(5)=1001
20       FOR J=1 TO 9
22       FOR I=1 TO 5000
24       FOR K=1 TO 5
30       IF A(K)-N(K)/F**J < B(K) THEN 50
40       GO TO 60
50       L(K)=B(K)
55       GO TO 65
60       L(K)=A(K)-N(K)/F**J
```

```
65      IF A(K)+N(K)/F**J > N(K) THEN 80
70      GO TO 90
80      U(K)=N(K)-L(K)
85      GO TO 100
90      U(K)=A(K)+N(K)/F**J-L(K)
100     X(K)=INT(L(K)+RND(X)*U(K))
102     NEXT K
105     P=1914258
110     FOR V=1 TO 5
120     P=P-(X(V)-S(V))**2+X(3)*X(V)**.8
130     NEXT V
140     IF P > M THEN 160
150     GO TO 170
160     A(1)=X(1),A(2)=X(2),A(3)=X(3),A(4)=X(4),A(5)=X(5), M=P
170     NEXT I
175     PRINT A(1),A(2),A(3),A(4),A(5),M
180     NEXT J
190     PRINT A(1),A(2),A(3),A(4),A(5)
200     PRINT M
210     STOP
220     END
```

```
817            931   671   573   30
2.22537E+06
823            966   718   554   116
2.24073E+06
822            974   718   555   129
2.24096E+06
816            991   712   524   120
2.24188E+06
829            987   714   529   125
2.24201E+06
825            984   715   525   122
2.24203E+06
825            986   716   526   123
2.24204E+06
826            986   715   525   123
2.24204E+06
```

```
827            986   716   526   123
2.24204E+06
827            986   716   526   123
2.24204E+06
```

EXAMPLE 12.6

Now let us try a ten-variable problem. Maximize $P = 900{,}000{,}000 - (x_1 - 370)^2 - (x_2 - 535)^2 - (x_3 - 892)^2 - (x_4 - 312)^2 - (x_5 - 427)^2 - (x_6 - 243)^2 - (x_7 - 183)^2 - (x_8 - 873)^2 - (x_9 - 919)^2 - (x_{10} - 944)^2$ subject to $0 \leqslant x_i \leqslant 1000$.

Since there are $1.01004512 \times 10^{30}$ feasible answers to this integer programming problem, we will adjust our multistage program to allow for ten variables and let it run.

In line 22 we upped the individual runs from 5000 to 10,000 because we are trying to get ten "good" random numbers each time instead of five. This might help a little.

The resulting program and printout below yielded $x_1 = 370$, $x_2 = 535$, $x_3 = 891$, $x_4 = 312$, $x_5 = 427$, $x_6 = 243$, $x_7 = 183$, $x_8 = 873$, $x_9 = 920$, $x_{10} = 944$ with a corresponding $P = 899{,}999{,}998$.

```
5     DIM A(10),B(10),L(10),N(10),      U(10),X(10),P(10),S(10)
7     F=2
10    X=1
11    S(1)=370,S(2)=535,S(3)=892,S(4)=312,S(5)=427
12    S(6)=243,S(7)=183,S(8)=873,S(9)=919,S(10)=944
13    M=568676982
14    A(1)=500, A(2)=500, A(3)=500, A(4)=500, A(5)=500
15    A(6)=500, A(7)=500, A(8)=500, A(9)=500, A(10)=500
16    B(1)=0, B(2)=0, B(3)=0, B(4)=0 B(5)=0
17    B(6)=0, B(7)=0, B(8)=0, B(9)=0 B(10)=0
18    N(1)=1001, N(2)=1001, N(3)=1001, N(4)=1001, N(5)=1001
19    N(6)=1001, N(7)=1001, N(8)=1001, N(9)=1001, N(10)=1001
20    FOR J=1 TO 9
22    FOR I=1 TO 10000
24    FOR K=1 TO 10
30    IF A(K)-N(K)/F**J < B(K) THEN 50
40    GO TO 60
50    L(K)=B(K)
55    GO TO 65
60    L(K)=A(K)-N(K)/F**J
65    IF A(K)+N(K)/F**J > N(K) THEN 80
70    GO TO 90
```

```
80     U(K)=N(K)-L(K)
85     GO TO 100
90     U(K)=A(K)+N(K)/F**J-L(K)
100    X(K)=INT(L(K)+RND(X)*U(K))
102    NEXT K
105    P=900000000
110    FOR V=1 TO 10
120    P=P-ABS((X(V)-S(V))**3)
130    NEXT V
140    IF P > M THEN 160
150    GO TO 170
160    A(1)=X(1),A(2)=X(2),A(3)=X(3),A(4)=X(4),A(5)=X(5)
161    A(6)=X(6),A(7)=X(7),A(8)=X(8),A(9)=X(9),A(10)=X(10)
162    M=P
170    NEXT I
175    PRINT A(1), A(2), A(3), A(4), A(5)
176    PRINT A(6), A(7), A(8), A(9), A(10)
177    PRINT M
180    NEXT J
190    PRINT A(1), A(2), A(3), A(4), A(5)
195    PRINT A(6), A(7), A(8), A(9), A(10)
200    PRINT M
210    STOP
220    END
```

The true optimum (from the factored form) is $x_1 = 370$, $x_2 = 535$, $x_3 = 892$, $x_4 = 312$, $x_5 = 427$, $x_6 = 243$, $x_7 = 183$, $x_8 = 873$, $x_9 = 919$, $x_{10} = 944$ with a corresponding P = 900,000,000. So we came pretty close for about one minute of computer time.

```
342         425   908   372   373
365         225   979   959   917
895104853
423         588   870   348   447
205         167   956   979   882
8985518959
391         556   882   332   392
247         159   913   906   945
899849517
```

344 528 884 326 443
226 185 875 924 966
899959027
368 537 897 323 431
239 184 872 915 930
899995590
368 538 887 308 428
246 187 874 916 946
899999648
371 536 894 313 429
246 183 875 920 943
899999944
369 536 893 312 427
244 183 874 918 944
899999994
370 535 891 312 427
243 183 873 920 944
899999998
370 535 891 312 427
243 183 873 920 944
899999998

A Multistage Monte Carlo Optimization Problem with Constraints

We have just looked at several multistage programs which optimized a number of different nonlinear functions. The author has run several more of these nonlinear optimization problems. A twenty-variable nonlinear function with 1×10^{61} answers was maximized. A thirty-variable nonlinear function was maximized; it had 1×10^{91} answers. In both programs less than 10,000 samples were drawn. For example, a thousand samples would be drawn throughout the entire n dimensional region. (20 dimensional or 30 dimensional region with each variable x being bounded as follows: $0 \leqslant x_i \leqslant 1000$ in these cases.) Then the n dimensional region would be shrunk by a factor of 2 in each dimension, using the "optimal solution so far" as the center of the n dimensional "rectangle." Then another 1000 samples were drawn in the smaller "rectangle." Each time a better answer was produced, the "rectangle" was recentered about the "new current optimal solution." After another 1000 samples, the region was reduced again by another factor of two and centered about the new optimum. This process continued until the rectangles (which are completely free to move anywhere throughout the n dimensional space at any particular time) funneled down to the right answer. The truly amazing features of this approach are that the approach always works (up to at least 100 variables), it requires very few samples, and is not too difficult as regards programming. Therefore, very little computer time is required.

This approach also works on any nonlinear* problem with an array of linear and/or nonlinear constraints. Again we use "rectangles" of every decreasing size that race through the n dimensional space (always staying inside the constraints) until the rectangles funnel into the right answer.

EXAMPLE 12.7

Maximize $P = -.02x_1{}^{2.2} + 1600x_1 + 8x_2{}^2 - .001x_1x_2 + 73x_3{}^{1.4} - x_3x_4 + 41x_4{}^{.5} + 52x_5{}^{.3} + x_6x_7{}^{.02} - 2.6x_8{}^{.3} + .0001x_3x_6 + 200{,}000{,}000 - (x_2 - 800)^2 - (x_4 - 600)^2 - (x_6 - 700)^2 - (x_8 - 900)^2$ subject to $x_i \geqslant 0$ for all i, and $x_1 + 2x_2 + 7x_3 + 2x_4 + 6x_5 + x_6 + 2x_7 + 5x_8 \leqslant 25{,}000$, $2x_1 + x_2 + 3x_3 + x_4 + 5x_5 + x_6 + 6x_7 + x_8 \leqslant 22{,}000$, $5x_1 + 10x_2 + 15x_3 + 20x_4 + 19x_5 + 4x_6 + 5x_7 + 6x_8 \leqslant 100{,}000$, $3x_1 + 3x_2 + x_3 + 2x_4 + x_5 + 10x_6 + 2x_7 + 5x_8 \leqslant 32{,}000$, and $8x_1 + 7x_2 + 6x_3 + 14x_4 + 6x_5 + 8x_6 + 8x_7 + 9x_8 \leqslant 80{,}000$.

The program and the printouts from two runs follow.

```
5          REM NONLINEAR MULTI STAGE MONTE CARLO
6          REM OPTIMIZATION WITH CONSTRAINTS
7          DIM X(8),C(8),D(8),E(8),F(8),G(8)
8          DIM A(8),B(8),L(8),N(8),U(8),S(8),T(5)
9          DIM H(8)
10         X=1
15         F=2
```

*Naturally this approach also works on linear problems. But because nonlinear problems "had" been considered more difficult and most applications are nonlinear, we will present nonlinear problems here.

```
16   M=0
20   FOR W=1 TO 8
25   B(W)=0
35   NEXT W
37   FOR W1=1 TO 8
39   READ N(W1)
40   A(W1)=INT(N(W1)/2)
41   NEXT W1
42   FOR W2=1 TO 8
44   READ C(W2)
46   NEXT W2
48   FOR W3=1 TO 8
50   READ D(W3)
52   NEXT W3
54   FOR W4=1 TO 8
56   READ E(W4)
58   NEXT W4
60   FOR W5=1 TO 8
62   READ F(W5)
64   NEXT W5
66   FOR W6=1 TO 8
68   READ G(W6)
70   NEXT W6
75   FOR R1=1 TO 5
80   READ T(R1)
85   NEXT R1
88   FOR R9=1 TO 8
90   READ H(R9)
92   NEXT R9
120  FOR J=1 TO 13
122  IF J >< 13 THEN 127
124  Z=10000
126  GO TO 129
127  Z=3000
129  FOR I=1 TO Z
132  FOR K=1 TO 8
230  IF A(K)-N(K)/H(K)**J < B(K)  THEN 250
240  GO TO 260
250  L(K)=B(K)
255  GO TO 265
260  L(K)=A(K)-N(K)/H(K)**J
265  IF A(K)+N(K)/H(K)**J > N(K) THEN 280
270  GO TO 290
280  U(K)=N(K)-L(K)
285  GO TO 300
290  U(K)=A(K)+N(K)/H(K)**J-L(K)
300  X(K)=INT(L(K)+RND(X)*U(K)+.5)
305  NEXT K
310  FOR W7=1 TO 5
```

```
320        S(W7)=0
330        NEXT W7
340        FOR W8=1 TO 8
350        S(1)=S(1)+C(W8)*X(W8)
360        S(2)=S(2)+D(W8)*X(W8)
370        S(3)=S(3)+E(W8)*X(W8)
380        S(4)=S(4)+F(W8)*X(W8)
390        S(5)=S(5)+G(W8)*X(W8)
400        NEXT W8
410        IF S(1) > T(1)  THEN 470
420        IF S(2) > T(2)  THEN 470
430        IF S(3) > T(3)  THEN 470
440        IF S(4) > T(4)  THEN 470
450        IF S(5) > T(5)  THEN 470
460        GO TO 560
470        B1=0
480        FOR W9=1 TO 5
490        IF S(W9)/T(W9) > B1 THEN 510
500        GO TO 520
510        B1=S(W9)/T(W9)
520        NEXT W9
530        FOR R2=1 TO 8
540        X(R2)=INT(X(R2)/B1)
550        NEXT R2
560        P1=-.02*X(1)**2.2+1600*X(1)+8*X(2)**2-.001*X(1)*X(2)
570        P2=73*X(3)**1.4
580        P3=-X(3)*X(4)+41*X(4)**.5+52*X(5)**.3+X(6)*X(7)**.02
582        P4=-2.6*X(8)**.3
585        P5=.0001*X(3)*X(6)+200000000-(X(2)-800)**2
587        P6=-(X(4)-600)**2
588        P7=-(X(6)-700)**2-(X(8)-900)**2
610        P=P1+P2+P3+P4+P5+P6+P7
620        IF P > M THEN 640
630        CO TO 670
640        A(1)=X(1),A(2)=X(2),A(3)=X(3),A(4)=X(4)
650        A(5)=X(5),A(6)=X(6),A(7)=X(7),A(8)=X(8)
660        M=P
670        NEXT I
680        PRINT A(1),A(2),A(3),A(4)
690        PRINT A(5),A(6),A(7),A(8)
700        PRINT M
710        NEXT J
1500       DATA 10000,10000,3571,5000
1510       DATA 4166,3200,3666,5000
1520       DATA 1,2,7,2,6,1,2,5
1530       DATA 2,1,3,1,5,1,6,1
1540       DATA 5,10,15,20,19,4,5,6
1550       DATA 3,3,1,2,1,10,2,5
1560       DATA 8,7,6,14,6,8,8,9
```

```
1570        DATA 25000,22000,100000,32000,80000
1575        DATA 2,2,1.82,1.9
1576        DATA 1.85,1.8,1.83,1.9
1580        STOP
1590        END
```

The First Printout is

```
2365
8
4.86996E+08
426
154
5.44109E+08
215
54
7.34873E+08
7
97
8.08546E+08
65
7
8.76214E+08
34
4
8.92692E+08
25
1
9.03947E+08
35
3
9.07006E+08
15
0
9.09798E+08
6
1
9.10913E+08
2
1
9.12714E+08
0
0
913558407
0
0
913700000
```

```
6265    551    2
36      161    613

6898    896    277
305     877    182

8638    93     133
2       319    374

9223    24     64
137     404    245

9728    10     18
143     8      122

9849    8      11
57      77     39

9930    10     7
13      4      32

9952    0      1
16      6      4

9972    2      1
15      6      6

9980    1      1
12      0      8

9993    0      1
0       0      0

9999    0      0
0       2      0

10000   0      0
0       0      0
```

The Second Printout is

```
1127
68
4.07381E+08
530
218
6.25855E+08
197
92
6.60258E+08
37
35
8.35620E+08
6
2
8.59837E+08
154
15
8.82318E+08
29
13
9.01839E+08
5
0
9.06594E+08
8
5
9.08679E+08
3
2
9.12014E+08
5
0
9.12445E+08
1
0
9.13145E+08
0
0
913700000
```

```
5320    208    947
364     92     936

7693    96     145
87      1171   299

8004    248    138
150     1287   368

9429    58     11
139     126    134

9608    23     70
30      4      287

9773    5      9
14      49     73

9915    1      3
46      10     7

9949    1      7
32      9      21

9964    0      1
8       4      24

9988    0      1
3       2      1

9991    0      1
4       2      2

9996    0      0
3       1      3

10000   0      0
0       0      0
```

The answer on both runs was $x_1 = 0$, $x_2 = 10{,}000$, $x_3 = 0$, $x_4 = 0$, $x_5 = 0$, $x_6 = 0$, $x_7 = 0$, and $x_8 = 0$ with a corresponding maximum P of 913,700,000. This is definitely the correct answer because the chances of two completely independent runs which sampled from about 1 x 10^{20} answers coming up with exactly the same wrong answer are infinitesimally small.

Each printout has thirteen subprintouts (the last one being the right answer). Each successive subprintout represents the optimum from the ever decreasing "rectangles." Again these rectangles are completely free to move in any direction across n dimensional space as far as they want to go in pursuit of the true optimal solution at any particular time. Notice how with each subprintout the answers get better and better until the last printout produces the correct answer. The first twelve subprintouts were the optimums of 3000 sample answers. The thirteenth one took a sample of 10,000 answers. Therefore, a total of 46,000 samples produced the true optimum even though there were many trillions times that many answers.

This program is conceptually the same as earlier ones in this chapter. There are only two extra features. First, from dividing coefficients in the constraints into the bounds on the constraints we learn that

$$
\begin{array}{ll}
0 \leqslant x_1 \leqslant 10{,}000 & 0 \leqslant x_5 \leqslant 4166 \\
0 \leqslant x_2 \leqslant 10{,}000 & 0 \leqslant x_6 \leqslant 3200 \\
0 \leqslant x_3 \leqslant 3571 & 0 \leqslant x_7 \leqslant 3666 \\
0 \leqslant x_4 \leqslant 5000 & 0 \leqslant x_8 \leqslant 5000
\end{array}
$$

Noticing that $2^{13} = 8192$, we choose 2 for the shrinking factor for x_1 and x_2. Because the limits are smaller on x_3 through x_8, we select shrinking factors of 1.82, 1.9, 1.85, 1.8, 1.83, and 1.9, respectively, for these variables. These values are assigned to the H(8) subscripted variables. Therefore, after thirteen runs lines 230, 260, 265, and 290 should guarantee rectangles with very small dimensions (only 2 or 3 units wide) around the true optimum, for the final (thirteenth) random search for the optimum. With the last pass looking at 10,000 samples, optimality is virtually guaranteed. And one can always run the program over again to be sure. (If we look at any one of the sampling distributions of optimization problems in the book, essentially what happens is that the first subprintout puts us in the upper, say, .05 region of the distribution. The second printout puts us in the upper .01 region, and then .001, .0001, . . . until the thirteenth pass gives us the right answer.)

The second extra feature is that the random numbers (8 each time) are read in and lines 310 through 450 check the five constraints. If all the constraints hold, then the program goes to line 560 where the function is evaluated and checked to see if it is the optimum so far. If it is, it goes to storage (lines 640 through 660). However, if the constraints do not hold, then the random numbers are sent to line 470 where lines 470 through 550 shrink them back inside the constrained region. Then the function is evaluated at lines 560 through 610 and checked to see if it is the optimum so far (storing it if it is).

Functions of 20, 50, or 100 variables should be easy to optimize with these techniques. (This program uses about 40 seconds of computer time on a fairly good sized computer.) Therefore, we can solve any nonlinear (and, of course, linear if we desire) optimization problem with or without constraints (and the constraints can be nonlinear, too) for up to 100 variables and even beyond. The "rectangles"[1] just move across the n dimensional space and go right for the optimal solution.[2]

The possibilities for new, more sophisticated and more accurate physics, engineering, econometric, business, financial, chemical, and medical models are virtually endless and very exciting.

Solving Equations with Multistage Monte Carlo

Certainly any nonlinear optimization problem of 50 or 100 variables is solvable with multi stage. But consider that equations and systems of equations can be treated as optimization problems, too, by using multistage Monte Carlo. Suppose we have the *nonlinear* system of equations:

$$f_1(x_1, x_2, \ldots x_n) = c_1$$
$$f_2(x_1, x_2, \ldots x_n) = c_2 \text{ and}$$
$$f_3(x_1, x_2, \ldots x_n) = c_3$$

We rewrite this nonlinear system as minimize

$$g(x_1, \ldots x_n) = |f_1(x_1, \ldots x_n) - c_1|$$
$$+ |f_2(x_1, \ldots x_n) - c_2| + |f_3(x_1, \ldots x_n) - c_3|$$

over whatever region of n dimensional space that you wish to look for solutions. Then you set up the multi stage program (usually 50-70 statements in Fortran or BASIC) and run it. A few seconds or a few minutes later, it finds $x_1, x_2, \ldots x_n$ such that $g(x_1, x_2, \ldots x_n) = 0$. This is a

1 By a "rectangle" in n dimensional space we mean an n dimensional box (sides at right angles in all dimensions, so to speak).

2 Note: If for some reason a "rectangle" moves in too fast and misses the optimal solution (at any of the 13 passes), the program is designed to notice this and slide the "rectangle" around until it covers the optimal solution again.

solution to the nonlinear system of equations and rerunning the multi stage program will produce new solutions (unless the solution is unique).

Using this approach, I have produced a solution to a one thousand variable nonlinear equation by only drawing a total of 3,500 samples as the one thousand dimensional rectangles rocketed to the solution. Each variable value was accurate to six decimal places, and of course the program automatically checks the answer to see that it is a solution. Run time is about two minutes on a large computer.

As an example, notice the forty variable nonlinear system of equations and one of its solution printed here. Polynomial equations with real and complex coefficients and roots are now solvable, as never before.

Write a multi stage program to find a solution to

$$\sum_{i=1}^{40} x_i^3 = 40{,}000$$

$$\sum_{i=1}^{39} x_i x_{i+1}^2 = 41{,}000$$

$$\sum_{i=1}^{39} x_i^2 x_{i+1} = 42{,}000$$

$$\sum_{i=1}^{38} x_i x_{i+1} x_{i+2} = 44{,}000$$

in the region $-25 \leqslant x_i \leqslant 25$ for $i = 1, 2, \ldots 40$.

The multi stage program is

```
2    DIM X(40),A(40),L(40),B(40),U(40),N(40)
4    X=1
6    M=1.0E30
8    F=4
20   FOR T1=1  TO 40
22   B(T1)=-25
24   A(T1)=0
26   N(T1)=25
28   NEXT T1
34   FOR J=1  TO  14
40   FOR I=1  TO 1000
45   FOR K=1  TO 40
47   IF A(K)-N(K)/F**J  < B(K) THEN 50
49   GO TO 60
50   L(K)=B(K)
55   GO TO 65
60   L(K)=A(K)-N(K)/F**J
65   IF A(K)+N(K)/F**J  > N(K)  THEN 80
70   GO TO 90
80   U(K)=N(K)-L(K)
85   GO TO 100
```

```
90    U(K)=A(K)+N(K)/F**J-L(K)
100   X(K)=L(K)+RND(X)*U(K)
102   NEXT K
110   S1=0
120   FOR J1=1  TO 40
130   S1=S1+X(J1)**3
140   NEXT J1
150   P1=ABS(S1-40000)
160   S2=0
170   FOR J2=1  TO 39
180   S2=S2+X(J2)*X(J2+1)**2
190   NEXT J2
200   P2=ABS(S2-41000)
210   S3=0
220   FOR  J3=1  TO 39
230   S3=S3+X(J3)**2*X(J3+1)
240   NEXT J3
250   P3=ABS(S3-42000)
260   S4=0
270   FOR J4=1  TO 38
280   S4=S4+X(J4)*X(J4+1)*X(J4+2)
290   NEXT J4
300   P4=ABS(S4-44000)
350   P=P1+P2+P3+P4
510   IF P <  M  THEN 530
520   GO TO 600
530   FOR R5=1  TO 40
540   A(R5)=X(R5)
550   NEXT R5
560   M=P
600   NEXT I
610   NEXT J
620   FOR R6=1  TO 36  STEP 5
630   PRINT A(R6),A(R6+1),A(R6+2),A)R6+3),A(R6+4)
640   NEXT R6
650   PRINT M
660   STOP
680   END
```

```
-12.4511       8.59451     -5.65879     -18.1396     -.860025
 17.7944       2.79516     -1.34553      20.4958     9.79884
-2.02146      -4.15157     -7.48315      15.5355     24.8157
 22.3542      24.3759       2.83235     -11.0386     -4.14937
 4.18555     -10.4186      10.1834      -5.69707     9.81483
 17.2441     -22.5377     -13.8703      23.0078      15.7571
 23.6558     -19.6128     -24.4615      14.1017     -21.6976
 15.9983     -22.0315      18.5563       5.13730    -12.5691
 8.31295E-05
```

The variable values are in the pattern

x_1	x_2	x_3	x_4	x_5
x_6	x_7	x_8	x_9	x_{10}
.	.	.	.	.
.	.	.	.	.
.	.	.	.	.
.	.	.	.	.
.	.	.	.	.
x_{36}	x_{37}	x_{38}	x_{39}	x_{40}

g.

Notice the g value is .0000831295. So this answer is accurate to five or six decimals in every variable.

EXERCISES

12.1 Maximize $P = 2x_1{}^2 + 140x_1 - 6x_2{}^2 + 992x_3 - 7x_3 + 1000x_2 - x_4{}^2 + 714x_4 - 2.2x_5{}^2 + 800x_5 - x_1x_2 + x_1x_3 - x_1x_4 + x_1x_5 + 7{,}000{,}000 - .1x_2x_3$ subject to $0 \leqslant x_i \leqslant 1000$ for $i = 5$.

12.2 Minimize P in 12.1 over $100 \leqslant x_i \leqslant 900$.

12.3 Maximize $P = -3x_1{}^2 + 590x_1 - x_2{}^2 + 1111x_2 + x_3{}^2 - 1125x_3 - 6x_4{}^2 + 731x_4 - x_1x_2 + x_1x_2x_3 - .001x_1x_2x_4$ subject to $0 \leqslant x_i \leqslant 500{,}000$ for $i = 1, 4$.

12.4 Maximize $P = -8x_1{}^2 + 2000x_1 - 7x_2{}^2 + 6000x_2 - 12x_3^2 + 40{,}000x_3 - x_4{}^2 + 85{,}000x_4 + 3x_5{}^2 - 17{,}000x_5 - x_6{}^2 + 12{,}000x_6 + x_1x_3x_5 - 17x_1x_6 + .72x_1x_5 - x_3x_4x_2$ subject to $0 \leqslant x_i \leqslant 1{,}000{,}000$, $0 \leqslant x_2 \leqslant 500{,}000$, $0 \leqslant x_3 \leqslant 400{,}000$, $200{,}000 \leqslant x_4 \leqslant 300{,}000$, $75{,}000 \leqslant x_5 \leqslant 100{,}000$, and $500{,}000 \leqslant x_6 \leqslant 1{,}000{,}000$.

12.5 Write a BASIC multi stage program to find the four roots of $x^4 - 15x^3 + 77x^2 - 153x + 90$.

12.6 Write a BASIC multi stage program to produce 10 solutions to the system of equations $\sum_{i=1}^{8} x_i^3 = 8{,}200$, $\sum_{i=1}^{7} x_i x_{i+1}^2 = 8{,}600$, $\sum_{i=1}^{6} x_i x_{i+1} x_{i+2} = 9{,}000$.

12.7 Modify the program in 12.6 to find the solutions in the region $0 \leqslant x_i \leqslant 28$.

12.8 Write a BASIC multi stage program to produce 15 solutions to the system of equations

$$\sum_{i=1}^{19} x_i x_{i+1} = 203{,}262 \text{ and } \sum_{i=1}^{20} x_i^2 = 202{,}428$$

in the region $-150 \leqslant x_i \leqslant 150$.

Suggested Reading

Conley, William. *Computer Optimization Techniques.* Princeton, N.J.: Petrocelli Books, 1979.

Conley, William. *Six selected articles.* International Journal of Mathematical Education in Science and Technology (1981).

Hadley, G. *Nonlinear Programming.* Reading, Mass.: Addison-Wesley, 1964.

Kemeny, John. *Man and the Computer.* New York: Scribners, 1972.

Orchard-Hays, W. *Advanced Linear Programming Computing Techniques.* New York: McGraw-Hill, 1968.

APPENDIX A

Statistical Forecasting

Throughout the text we have used various forecasted, predictive, or regression models to develop some of the functions that we want to optimize. Let us look at a few of the techniques for calculating the forecasted equations.

Straight line between two points

This technique can be used to develop the downward sloping demand curves we used in the book. Suppose your boss claims that 50,000 units of your product can be sold if it is priced at $100 per unit and 80,000 units can be sold at a price of $90 per unit. Therefore, we want to find the straight line connecting these two points. The general equation for a straight line is $y = mx + b$, where m is the slope and b is the y-intercept. The two points $(x_1, y_1) = (50{,}000{,}100)$ and $(x_2, y_2) = (80{,}000{,}90)$ give the line's slope $m = (y_2 - y_1)/(x_2 - x_1) = (90 - 100)/(80{,}000 - 50{,}000) = -.000333333$. So the line is $y = -.000333333x + b$. Substituting $(x_1, y_1) = (50{,}000{,}100)$ into the line yields $100 = -.000333333(50{,}000) + b$ or $100 = -.16.66666 + b$ so $b = 116.66666$. The line is $y = -.000333333x + 116.6666$. And this line could be used as the price-quantity curve.

If more than two points were used for estimates, then the straight lines connecting the adjacent points could be used and we would have a piecewise linear demand curve that the computer could deal with easily.

Straight line best fitting several points

To find the "best" straight line through a series of points, solve for the slope using the equation $m = (n\Sigma xy - (\Sigma x)(\Sigma y))/(n\Sigma x^2 - (\Sigma x)(\Sigma x))$, where n is the number of ordered pairs and Σ means summation. Then solve for the y intercept b with $b = \overline{y} - m\overline{x}$, where y means the numerical average of the y values and x means the numerical average of the x values.

As an example, find the best straight line through the following points:

x	y
1	5
2	8
3	11
4	15

Setting up a chart gives us:

x^2	x	y	xy
1	1	5	5
4	2	8	16
9	3	11	33
16	4	15	60
$\Sigma x^2 = 30$	$\Sigma x = 10$	$\Sigma y = 39$	$\Sigma xy = 114$

So $m = (4 \text{ x } 114 - 10 \text{ x } 39)/(4 \text{ x } 30 - 10 \text{ x } 10) = 66/20 = 3.3$. Therefore, $b = \overline{y} - m\overline{x} = 39/4 - 3.3 \text{ x } 10/4 = 9.75 - 8.25 = 1.5$. The straight line that is "closest" to the data points is $y = 3.3x + 1.5$.

In this case the line is very close to the data points, so if this represented some type of predictive model, the line would be good for prediction. But this is not always the case. Use of the formulas for m and b always gives the best straight line, whether the fit is good or bad. When there is doubt about the relationship being investigated (or the strength of the relationship), then the line should be tested statistically before much faith is placed in it for use in an optimization model or anything else. (Any statistics book will discuss the simple t test for the "goodness of fit" of the fitted straight line.)

Multiple variable forecasting

Let us say that we are trying to model sales y as a function of price x_1 and number of salespeople x_2 in each area. The data from various sales areas (of equal sales potential) over the past few years has yielded the following:

x_1 price in cents in the area	x_2 no. of sales-people in the area	y sales in the area in units of millions
52	12	58
49	12	66
51	14	60
48	10	57
45	10	72
45	11	74

Therefore, we have:

$x^2{}_1$	x_1	x_1x_2	x_2	$x_2{}^2$
2704	52	624	12	144
2401	49	588	12	144
2601	51	714	14	196
2304	48	480	10	100
2025	45	450	10	100
2025	45	495	11	121
$\Sigma x_1{}^2$ is 14,060	Σx_1 is 290	Σx_1x_2 is 3,351	Σx_2 is 69	$\Sigma x_2{}^2$ is 805

y	x_1y	x_2y
58	3016	696
66	3234	792
60	3060	840
57	2736	570
72	3240	720
74	3330	814
Σy is 387	Σx_1y is 18,616	Σx_2y is 4,432

Now if we try to fit the model $y = B_0 + B_1x_1 + B_2x_2$, then we have

$$nB_0 + B_1\Sigma x_1 + B_2\Sigma x_2 = \Sigma y$$
$$B_0\Sigma x_1 + B_1\Sigma x_1^2 + B_2\Sigma x_1x_2 = \Sigma x_1y$$
$$B_0\Sigma x_2 + B_1\Sigma x_1x_2 + B_2\Sigma x_2^2 = \Sigma x_2y$$

Note: x_2 could equal x_1^2 if we had just one variable and wanted to fit the best parabola $y = B_0 + B_1x_1 + B_2x_1^2$.

Substitution from our chart yields:

$$6B_0 + 290B_1 + 69B_2 = 387$$
$$290B_0 + 14{,}060B_1 + 3351B_2 = 18{,}616$$
$$69B_0 + 3351B_1 + 805B_2 = 4432$$

So solving this 3 by 3 system of equations will yield B_0, B_1, and B_2 that produce the best line of the type $y = B_0 + B_1x_1 + B_2x_2$ for the data. Again, then, this line might have to be tested statistically to see if it is good for prediction.

Also, perhaps a model of the type $y = B_0 + B_1x_1 + B_2x_2 + B_3x_1x_2$ would be better for the data. Or perhaps $y = B_0 + B_1x_1 + B_2x_2 + B_3x_1^2 + B_4x_2^2 + B_5x_1x_2$ would be better. These can all be created and tested through similar procedures. (See a statistics book for details.) At some point the data and subsequent model produced from solving the attendant system of equations become so cumbersome to calculate that a computer should be employed to do the work. Computer routines have been written to do the statistical work, solve the system of equations, and test the model for "usefulness" statistically. These computer routines should be employed in large forecasting studies where possible.

These models, of course, can be used on business data, econometric forecasts, chemical yield data, etc.

Suggested Reading

Bowerman, B.L. and O.Connell, R.T. *Forecasting and Time Series.* North Scituate, Mass.: Duxbury Press, 1979.

Kleinbaum, D. and Kupper, L. *Applied Regression Analysis and Other Multivariate Methods.* North Scituate, Mass.: Duxbury Press, 1978.

APPENDIX B

The Philosophy of Mathematics

Calculus optimization techniques have been useful in solving "least squares" equations with n variables, giving rise to statistical forecasting work. This is a classic example of the power of calculus. However, as the speed, capacity and availability of computers increases, it can be clearly seen that for any practical optimization problem of one hundred variables or less, calculus should be abandoned and a computer program written to optimize the function.

The most powerful of these computer optimization techniques is the multistage Monte Carlo integer program. This program runs repeated random searches for the optimum. Since after each run the computer focuses more closely on the region containing the true optimum, the optimum can be found quickly.

Calculus was and still is one of the greatest scientific discoveries in history. But our reverence for calculus is holding us back and preventing us from solving practical nonlinear optimization problems. Calculus was invented because people were unable to do calculation accurately for long periods of time. Therefore, every piece of mathematical philosophy had to be based on this fact. All theories had to be designed so that one could throw out the millions of answers and get right to the correct answer with just a little calculation. But today we are not burdened with these limitations, and we should devote our mental efforts to "asking the right question" and let the computer find the solution. Whether this solution takes ten calculations or ten million calculations, really does not matter to a computer.

One can realistically ask whether calculus would have been discovered and developed at all if computers had been invented in, say, the fifteenth century. We are still so close historically to the discovery of computers that it is difficult to judge their final impact on mathematics.

Another problem in mathematics is a preoccupation with variables and writing things down functionally (again, where this is appropriate, it should be done). Consider a site location problem involving 100 cities. The objective would be to select the city that minimizes the sum of the distances from that city to all the other cities. A mathematician might say that this is a function of 10,000 variables subject to dozens of constraints. This is, of course, true, but a computer programmer can see that all that would be necessary would be to add up the mileages (to the other cities) for each city and take the minimum. This would require about one-tenth of a second of computer time.

Again there are and will continue to be many and varied applications of theoretical mathematics, where variables, limits, derivatives, integrals, continuity, and other concepts are important. Calculus should continue to be taught. There are engineering applications where the integral of the cosine equaling the sine is still important, regardless of computer developments.

As mathematics move into perhaps its most exciting era (in terms of the development of ways to solve so many problems), it will need the full participation of its theoretical and computer people.

Index